MARS

OUR FUTURE ON
THE RED PLANET

M A

BOOKS BY ROBERT M. POWERS

Planetary Encounters

Shuttle: World's First Spaceship

The Coattails of God

Other Worlds Than Ours

Mars: Our Future on the Red Planet

R S

ROBERT M. POWERS

HOUGHTON MIFFLIN COMPANY

Boston   1986

Library of Congress Cataloging in Publication Data

Powers, Robert M., date
    Mars: our future on the red planet.

    Bibliography: p.
    Includes index.
    1. Mars (Planet) — Exploration.   2. Life on other planets.   I. Title.
QB641.P73   1986   919.9'2304   85-30534
ISBN 0-395-35371-8

Printed in the United States of America

S 10 9 8 7 6 5 4 3 2 1

To Percival Lowell, for the dream

CONTENTS

MARS

PROLOGUE

AMONG THE MORE scientifically nonsensical expressions of the human race is "It's in his blood." This facile remark, which takes little notice of psychiatry, medicine, or the periodic table of the elements and no doubt history, anthropology, or the principles of Archimedes either, persists among us. It has done so since the periodic table of the elements consisted of earth, air, fire, and water.

The "It's in the blood" explanation has been applied to third-generation military people, doctors, lawyers, artists, and scientists. Jean Dominique Cassini, the first director of the Paris Observatory, founded a dynasty of astronomers who worked from 1700 to 1800, during which time there was always a Cassini in some position of importance in the world of star gazing. In the end, one finally deserted to be a botanist.

But what if someone claims to be a Martian — or at least professes a deep affinity for the Red Planet and persists in a belief that he or she might once have lived there? Would we invoke "It's in the blood" to explain it?

On a reasonably nice day in Los Angeles in January 1979, I was on my way to Beverly Hills for lunch with Ray Bradbury. I was on assignment for a fashionable current events magazine in the east, whose editors I had persuaded that the scientific investigation of Mars was at least as important as running with the illegal-immigrant border circuit in northern Mexico or the antics of the genuine characters who inhabit the rodeo world. I was carrying with me a copy of *Mars and the Mind of Man*, a 1973 book containing comments and speculations about the photography of Mars by the *Mariner 9* spacecraft in 1971. It was written at a time when an inhabited Mars was a particularly weak suggestion, in view of the rather bleak photos of the planet so recently transmitted.

In the book, Ray Bradbury had written that he had always looked on himself as a Martian; indeed, that he amiably chose to believe that he had once lived on Mars, if one considered reincarnation a viable possibility. He had gone on to explain that a certain Dr. Electrico had informed him they had met on a battlefield in the Argonne in France during World War I, where he had died to be retreaded as Ray Bradbury. And if he could have died in the Argonne and come back, then he could just as easily have lived on Mars and come back.

Lunch in this case was down a back street in a sidewalk café that faced west. Ray arrived by taxi a few minutes after I had parked whatever clunker I was driving that year amid the Rolls-Royces, Bentleys, Mercedes, and assorted other finery that lived in a parking garage down the street. We took a table in the sun near the front door, and after a lengthy interval the menus arrived.

In honor of the occasion, I aimed for a Ray Bradbury

Burger, but the waitress spied me as a *turista* and let me know it was steak tartare. In the dim corridors of memory, I no longer recall what I ate, nor what Ray ordered, but I do remember the conversation. It was about Mars, naturally. I said it looked a bleak planet from the photos. He said it was lonelier than Barsoom ever was; it was still beautiful, though, still perhaps full of life, even if nothing more alive than the ghosts of the Martians. I said we would live there in the times of my daughter, or perhaps my daughter's daughter. He said, "Oh, no, much sooner than that." He didn't ask how old my daughter was.

We spoke of a short story he had written nearly thirty years before, "Dark They Were and Golden Eyed." I had read it in an anthology while still in grammar school. In the story, a man and his family help colonize Mars. They eat the foods and live in the strange seasons, experience the savage conditions on our first planetary outpost. The rest of the colonists leave, much in the way the Vikings finally left Greenland when it became intolerable to maintain colonies. But the family stays, and over the years their physical aspect changes: their flesh melts into new shapes; their skins assume new tints; their eyes, once the eyes of Earth, become flecked with gold. Eventually they move up into the hills and live amidst the ruins. They have become Martians.

It is a typical Ray Bradbury story, full of humanness and romance, the wonder of Mars, and the palpable presence of the ubiquitous old ones of the planet. He could have turned this somewhat elegant view and writing style upon any subject, upon something more easily grasped in human experience. But he chose Mars.

The sun was headed toward the beaches when we left the table. Whether he was recognized by anyone in the café I do not know. As he headed for a taxi by the curb, he squinted at the sun. "I love that planet," he said. And in the flyleaf of my book he had written, "For Bob, who loves Mars almost more than I do." On the way to my hotel, I wondered about this business of loving a planet. Clearly Ray saw nothing odd there. Nor did I.

What is it we love about a painting? A sculpture? An automobile as art, like a Bugatti or Ferrari? Talk to a Russian about Mother Russia, and he's not speaking of politics. He's talking about the land itself. These are all inanimate, and yet people profess to love them. Indeed, books are inanimate, too, and we love them, cherish them, reread them; we buy them at auctions, preserve them in expensive libraries.

I have had this discussion with many people. "How can you love something not alive?" they ask. "How can you equate the warmth of humanity with the swirl of colors on a flat canvas, the cold of marble, the silky texture of sculpted bronze, or the multi-decibel scream of a V-12 racecar at the red line? Can you love an old rare edition as you can a friend, a piece of eighteenth-century furniture as a brother; is your tiny piece of earth so precious that you can be said to care for it as a son or daughter? How can you love a planet, truly?"

People can come to understand that love more easily with a painting, a sculpture, an antique chair, a rare book, perhaps because those objects arouse our senses in some way, seem familiar to us, comprehensible. They see it less easily with a racecar and even more dimly with a planet, if at all. In any event, I am often told, it is a one-sided love, traditionally unhealthy. They miss the point. All of these inanimate objects are but ideas. They are reflections of the mind of a human behind them.

It is quite possible to love an idea, because there *can* be an interaction. You can feel the faint pulse of a thought over centuries, even millennia. You can grow close enough to it that you never want to see it die. You follow it with a whole sack of human emotions, committed to the love of a thoroughly inanimate object. So it is possible, I think, to love Mars not because it is a stone cold and lonely rock but because we can, and have, put warmth into it over the centuries. It is the human construction that we love.

A hundred astronomers have left parts of their souls and their hopes in drawings showing the surface of Mars. A score of men have left their stamp in the major theories about life on the strange planet fourth from the sun. The names of ten thousand technicians and scientists rest now on a plaque standing a few feet above the soil of Mars, attached to a spacecraft sent there in 1976. Fifty writers have tried their pen out on Mars and things Martian; sixty movie directors have tried to grasp the magic and mystery.

In these pages, I would like to show you how to fall in love with a planet.

1 FAR LONELIER THAN BARSOOM

THERE WAS A TIME when talking about landing on Mars and living there would generate a good deal of polite laughter. For some years such talk was confined to the pages of science fiction stories and Hollywood films, some more poorly crafted than others. When the first real on-paper mission plan came along, in 1947, the author was urged by publishers to make it an appendix to a novel, since the idea of it as a nonfiction book was too outrageous. Those days are long gone.

This book is about landing on Mars and living there. It is the story of a future human race — a not very distant one — which will set down on the terrain we have so often imagined in talk and stories and films. It is about Martians. Let me introduce them: they are young children now. They live on the planet Earth, as their forefathers did. They will not always do so. *Their* children will talk knowingly about changing an entire world to make it more like home.

If an unprotected human stood amid the rocky landscape

of Mars today, the result would be swift and brutal. Landing in Chryse Planitia, the Golden Plain of the Red Planet, the human's spaceship would settle down on the surface; reddish dust would swirl and billow beneath the rocket engines. A honeycomb aluminum ladder would protrude from a panel in the lower cargo bay and drop to the surface.

The sun would be much smaller than the warm, generally friendly one we know, because Mars is farther away from it. The wind would come from the east at 15 miles per hour. Around the landing site would be a scattering of rocks and boulders, some brown, some reddish-mustard color.

There would be sand in the air, tiny flakes of orange-red regolith, the "soil" of Mars. It would drift across the landscape, a thin veil occasionally obscuring the spongy, jumbled rocks that form the rim of an ancient, eroded crater.

The sky would be a bright salmon pink; to the west would be thin cirrus clouds tinged with pale rose. The temperature would be 60°F, high noon on a warm Martian summer's day. The tiny potato moon Phobos would be high in the sky, a small disk moving from east to west. It is one of the "hurtling moons" of Barsoom, described in fiction so long ago by Edgar Rice Burroughs.

Far away, beyond where the eye could see, beyond the closer than Earth-normal horizon, would lie the Great Tharsis Bulge, a huge goiter on the face of the planet. It covers a quarter of the globe. From the Tharsis Region rises Olympus Mons, nearly 17 miles above the average surface level — sea level, if a sea existed on the planet. The jagged rim of the volcano's caldera is covered with frost.

Looking through a glass port in the airlock door at the ladder descending to the ground, the human would press a switch and the door would open. In the weak light of the Martian day, the unprotected human would make the first step down the aluminum rungs. Death would be final and sharp.

In the first five seconds, the life-giving oxygen of Earth would be ripped from the lungs by the near vacuum that passes for an atmosphere. There would be no chance to hold that last breath; the rush of air from the lungs, high pressure of Earth to low pressure of Mars, is a gift of physics. It is an inexorable right of gases to flow this way, unchanging, unchangeable, and terminal. The ruptured lungs, hemorrhaging in a hundred capillaries, a dozen arteries and veins, would struggle vainly to complete the natural cycle the brain stem would still be broadcasting: breathe in, breathe out.

The only thing to breathe is a gritty mixture mainly composed of carbon dioxide. There would be only a trace of oxygen in that never-taken fatal breath. The brain would feel the ruptured lungs, sense the awful craving for life-giving oxygen, have one fleeting image of the sun and the limpid daylight.

The human would fall from the ladder, dropping to the ground slowly in the weak Martian gravity. The lower limbs would already be numb. The heart would stop, arrested in shock. Amid all the terrible sensations that would be experienced, the human would hardly notice a rapid warming of the skin from an ultraviolet ray bath much greater than on Earth. It would be the start of a savage sunburn, worse than any high-altitude sunburn on earth.

The human would lie in the coming dusk like any dead creature in the desert: curled in a fetal position, lips black, the body withered and faded. Lying on that ochre plain, the corpse would mummify, decaying extremely slowly; the Martian air is a thousand times more arid than the dry air of Luxor which preserved the kings of ancient Egypt.

Well, no one said it was Newport Beach. What has been

said is that Mars is more Earth-like than not. That is the conclusion we've come to after studying the planet for three and a half centuries by telescope and for the last two decades by robot spacecraft.

In the early hours of Tuesday morning, July 20, 1976, Jet Propulsion Laboratories looked like a giant offshore drilling rig highlighted against a black ocean. Nearly every building was alive. The sidewalks were illuminated by row upon row of lights. The glow from the complex could be seen from miles away along the Foothills Freeway. The parking lot was overflowing.

Seven hundred men and women at JPL had been involved day and night with a spacecraft for almost a year as they navigated it from Earth to Mars along a 505-million-mile flight path. The craft's name was *Viking 1,* and in a few hours it would land on Mars and photograph the surface. There was a second lander and also two orbiters composing the complex mission. Not many of the men and women in the buildings would have said with easy assurance that their grandchildren might one day inhabit the planet the little spacecraft was about to visit.

What we call Mars — not just the rocky planet, but the whole great dream — is a product of three fundamental human characteristics: philosophy, exploration, and conquest. There has never been a civilization, no matter how small, insignificant, or short-lived, whether with or without written records, that did not attempt to explain itself and provide some sort of intellectual groundwork for the question of why it existed. We call this search philosophy, and it is why scientists used to be called "natural philosophers."

Fundamental to elementary and complex ideas about existence is one of two possible alternatives: we are alone or we are not alone.

Speculation about something "out there" can be as simple as wondering whether there is life beyond the mouth of the cave or beyond yonder hill. It can be more complicated: is the next island inhabited, and by whom or what; does life exist anywhere across the Great Sea? Is there life on Mars?

This sort of human speculation has always played a strong role in cartography. A medieval map shows the earth as a circular disk with Jerusalem at the center. The map is not oriented north, as we would have it today, but east, that being the direction of paradise. Europe and Africa are in transposed positions, Spain has a triangular shape, Italy's boot is hardly identifiable, and England and Wales are shown as an elongated island with a narrow band of water dividing them from Scotland.

Myth and fantasy peopled the world. Not only were "we" not alone, but the companions on our planet included mermaids, unicorns, dragons, headless men in Africa, four-eyed Ethiopians, a race of men with only one foot, used like a parasol; there were horse-footed men, birds with feathers that glowed in the dark. There were ants as large as mastiffs, gryphons, and one-eyed hunters who drank mead from tankards made from their parents' skulls. Speculation without information can fall very short of reality.

You can start a fire just by rubbing two dry theories together, so it was left to the second fundamental human urge to provide the answers in the form of factual information. The wonderful ancient theories of geography, the mythical lands, the fantastic inhabitants of those distant places, all changed with the rise of exploration. To prove what *really* existed, some strange, tormented souls and their hostage crews, who had signed on because there was little other opportunity, sailed to the edges of the deep to discover wherein the truth lay. Most of the speculations — theories — were resolved over the next few centuries, the Age of Discovery. The explorers found that they were indeed not alone, that there were hundreds of civilizations out there.

Exploration was followed by conquest. Colonies were established in every country and land on the globe where humans could live in any comfort and where the residents, if any, could be encouraged to move out or tolerate the superimposition of a new culture on the indigenous one.

To dream, to explore, to colonize, those are the basics; and while it has taken a bit of time to extend that sequence outward to the planets (there have been as many misconceptions about the planets as in the medieval map of our own world), the nature of the game is still the same. Giordano Bruno, Bishops Godwin and Wilkins, de Fontenelle, Voltaire, Jules Verne, H. G. Wells, Edgar Rice Burroughs, Ray Bradbury — these were the philosophers and dreamers. The exploration was without ships or crews, without physical presence. Galileo, Huygens, Herschel, Schiaparelli, Flammarion, Lowell, and Antoniadi replaced the great sailors of the past. This voyage, one of the most challenging investigations of human history, proved that the planets were not just lights in the sky or gods; they were worlds. It was not a great leap of imagination to speculate on what they might be like and if they resembled our own — most particularly if they resembled our planet in having life.

An inhabited or habitable Mars has persisted in both the public and the scientific mind for centuries. This is the legacy of the Red Planet and is what has made it so special. Even people who cannot name more than two of the nine planets invariably identify Mars as one. In the seventeenth century we turned telescopes on it and saw what we took to be seas and continents. In the eighteenth century we learned there were polar caps. We also knew by then that the "day" on

Mars was similar to our own; it was slightly more than twenty-four hours long. Our crude instruments and our imagination told us Mars was warmer than it is, had more atmosphere than it does; we thought it had adequate, but dwindling, supplies of water. Thus was born our most precious fantasy: the planet was the home of an intelligent race that built a string of canals to pump water to the dying cities of the Martian deserts.

The twentieth-century telescopes and sophisticated scientific instruments made a shambles of the dreams. The canals weren't there. The planet was far colder than anyone imagined. The atmosphere was very thin and most probably devoid of life-giving oxygen, without which no life we know of could exist. The final blow came with twenty-two photographs by the little spacecraft called *Mariner 4*. They showed us a bleak, cratered terrain, a harsh, dead planet. Venus was already known as an acid sauna. Savage reality descended: we were truly alone in the solar system. Earth was a wonderful oasis between two worlds upon which life did not exist and upon which we could not live.

But four years after the flight of *Mariner 4*, NASA launched two further vehicles, *Mariners 6* and *7*. They confirmed the superficial resemblance of the Martian surface to the moon; but as the information from the spacecraft unfolded, scientists realized that Mars was not as much like the moon as they had thought in 1965. For one thing, there was a crater, 300 miles across, with frost on the rim. No frost had ever graced the rim of a lunar crater. Frost meant water.

The vast dark regions appeared to be a result of volcanic action during some distant geologic epoch. There was a cloud of haze over the southern polar region — it was winter there in 1969 — that seemed to be formed from ice particles and water vapor. The southern polar cap was estimated at several feet thick, not the thin rime crust that had previously been theorized. All this added up to a planet rather unlike the moon — more like Earth, in fact, as had always been supposed.

Although there were plenty of craters, they were not everywhere, as they are on the moon; many areas on Mars were completely free of them. The temperatures at the surface were a bit Antarctic but tolerable: +50°F at the equator at local noon, dropping to −153°F at night. The quick drop in temperature after sunset was a function of the very thin atmosphere. The spacecraft showed a "new" Mars that had water in the forms of vapor and ice (the atmospheric pressure was too low for liquid water), fairly decent temperatures, sunlight, soil, clouds, frost, and a little oxygen. Any life there might be rudimentary but could not be ruled out completely.

Fly-by missions, scientists decided, were not returning enough data in the short time over the planet to confirm or deny a hundred questions. The basic question was, How Earth-like is it? That could only be answered by an orbiting spacecraft, a Martian artificial satellite that could remain over the planet for months, even years if necessary — and by a landing vehicle. The former was called *Mariner 9,* the latter *Viking.*

On May 30, 1971, *Mariner 9* left Earth's orbit so accurately that only one midcourse burn was necessary to drop it into the correct orbital slot on November 13. The low point of the orbit was 862 miles above the surface of Mars. The spacecraft carried two sophisticated television cameras, one of which was a telephoto. But the first picture from the most expensive planetary mission that had ever been flown contained nothing. It was blank because of a Martian dust storm that amounted to a global hurricane. Estimates indicated winds of 100 miles per hour. The storm kept up for a month. The Martians, had there been any, would have been talking about the unseasonable weather. Few people argued any longer over whether Mars was a dead world. Those who said it was were invited to go there and enjoy the storm from the surface.

When the winds died in December, the cameras picked up faint details emerging like some fantastic rock born from the sea: the tops of massive volcanic mountains came into view. The great frost-rimed crater seen in 1969 turned out to be a volcanic caldera sitting on top of the largest volcano anyone could imagine. Long tongues of lava flows covered its slopes.

There were more surprises in store. Three other giant volcanoes showed up 700 miles away. Another group was spotted in an area called Elysium. The planetary geologists quickly concluded that vulcanism had been a part of Martian history between two billion and 200 million years ago. The largest volcano was the youngest in geologic time.

There was a great rift valley that they called Valles Marineris: 4 miles deep, 43 miles wide, and almost 3,000 miles long. There were adjoining valleys of a not inconsiderable extent; they looked like nothing less than water erosion channels. "Well, we found the canals," one scientist remarked. He was partly right. The linear features of Schiaparelli, Lowell, and others, stretching across the planet in an intricate network of oases and lakes, were dead issues. In many ways the old Martian canals weren't as interesting as the geological reality of the channels. Moreover, there was a gigantic network of what must once have been a river system.

There had been rain: the ice had melted, the permafrost had unfrozen, and gushers had poured forth to inundate the landscape. There were unmistakable signs of headwaters, tributaries, river valleys, islands, sandbars, and channels. The river system ran northward toward a basin that might once have been a sea.

Just *finding* all this was fascinating. But what it told the scientists was enormously important. Some of the river systems appeared much older than others, which meant that Mars had experienced more than one cycle of flooding and therefore had at times been very different. Life might have evolved on the planet in an earlier epoch and then adapted to the present conditions. No one was seriously proposing intelligent Martians, of course, but the idea that life of some kind lived there could not be easily dismissed.

This was why so many people from such diverse backgrounds gathered at JPL in July 1976. The importance of the *Viking* landing was immense, and the mission was a gamble. The little lander stood only a few feet above the ground on thin aluminum legs, and a medium-sized boulder could end the exploration before it began. There were plenty of boulders that size on Mars. There were also depressions. If the footpad of the lander got stuck in one of those, it might collapse, leaving the lander's cameras pointing helplessly at the sky with only the two Martian moons in its camera range, if the lander was tilted just right.

It was a historic moment for one reason above all others. One of the principal experiments aboard the lander was a little laboratory designed to find and define any Martian life in the soil or atmosphere. Scientists already knew that there was nothing on Earth's moon but regolith, the lunar "soil." Venus was a chemical hell of hideous temperatures — the Russians had proved that with their expensive and thorough investigations by the Venyera unmanned spacecraft. Mercury was a lunar-like body with craters, heat, and nothing else. The Big Planets — Saturn, Jupiter, Uranus, and Neptune — were giant gas balls incapable of sustaining any life we could think of.

It all came down to Mars, and at that moment it was

coming down on the slightly idiotic computer aboard a 10-foot-wide spaceship made of aluminum-titanium, spun fiber glass, and Dacron cloth. It was only 7 feet tall from its footpad to the top of its disk-shaped S-band antenna. It landed in a 1,000-mile-wide basin in the northern hemisphere. Almost 3,000 miles from the landing site was the great volcano Olympus Mons. About 2 miles away was the rim of an old, eroded impact crater. The landing was in a low area of the basin, where many of the biologists thought the conditions for life were reasonably good. Low areas on Mars have more atmospheric pressure and possibly some moisture.

It was "Sol 1," as the scientists referred to the Mars-based mission time. Line by line, the cameras traced out the first landscape photo. The 300-degree panorama was as clear as if it had been shot in a studio. That desert scene, except for the absence of vegetation, could have been in the Arizona-Sonora Desert.

The surface was more dark than light and the rocks were varied, both in size and contrast. The Martian soil was, not unexpectedly, rusty, orange-red. The sky was bright — a bright blue, the most Earth-like aspect of the photo. The blue sky lasted until JPL engineers calibrated the color images with the colors of an American flag attached to the spacecraft.

The real sky color turned out to be pink, caused by particles of rust-red dust in the atmosphere. In a long series of fantastic surprises Mars had in store for scientists, the public, and the press, the pink sky was the most astonishing. No one had ever dared suggest that the Martian sky might be pink, although in hindsight many scientists were sure they should have predicted it, based on information available before *Viking* left.

With the exception of the strange sky, the view from the surface was still strikingly Earth-like. No life of any kind could be seen within view of the lander cameras. But there are thousands of places on Earth where a photographer would not record life either, and the absence of visible life was not unexpected. For those who still hoped to find organisms there would be an interminable wait. But we had waited all this time for an answer to whether there was more than one inhabited planet in the solar system and we could perhaps wait a little longer. The biology experiments were not scheduled to begin until Sol 8 — July 28, 1976.

Meanwhile, *Viking* was busy supplying answers to hundreds of questions that had been debated for most of the twentieth century. The atmosphere was 95 percent carbon dioxide, 2–3 percent nitrogen, 1–2 percent argon, and 0.1–0.4 percent oxygen. The nitrogen, suspected but never before confirmed, was a very positive sign. Nitrogen is basic to life on Earth and is present in all living organisms on our planet. The atmospheric pressure at the landing site was not quite one percent of Earth's sea-level pressure.

If the atmosphere was not what we would consider a good breath of fresh air, the temperature was also not what we would call a typical summer afternoon, except perhaps at our South Pole. It was $-22°F$ and dropping. On July 21, Sol 1, the first weather report and forecast from another planet was made by Dr. Seymour L. Hess of Florida State University, based on information from *Viking Lander 1*: "Light winds from the east in the late afternoon changing to light winds from the southeast after midnight. Maximum wind was 15 miles per hour. Temperatures ranged from $-122°F$ just after dawn to $-22°F$." It was a grand day on the Red Planet.

On Sol 8, the soil sampling unit on the Mars lander was commanded to extend its arm — with shovel attached — and bring in samples for analysis. It did as ordered and delivered two small scoops of dirt into the biology experiment. The dirt turned out to be very rich in iron (14 percent) with varying amounts of calcium, silicon, titanium, and aluminum. On Sol 12, the experiment to detect microorganisms showed a rapid emission of oxygen from a small sample. Unfortunately, the process involved could be explained by chemical rather than biological reactions; and it was generally concluded that the success, such as it was, of the experiment did not confirm that life of any kind had been found. What it did show was that there was a very quick and easy way to get life-giving oxygen on the planet, and that had applications for any humans who might one day land there.

Another *Viking* biology experiment exploded with quick results when Martian soil was incubated in a nutrient the scientists called "chicken soup." Once again, unfortunately, it was shown that the result could just as well be chemical as biological in nature. That the stuff had some weird chemistry was not quite as surprising as finding biological life in the rather harsh Martian conditions would be. A further experiment gave confusing results perhaps indicative of biological reactions — and, as usual, perhaps not.

The data was analyzed and repeated and repeated and analyzed. No one wanted to be the first person to claim there really was life on Mars — and then be proved wrong. And no one wanted the albatross of being the one who said there wasn't any life and then have it proved beyond any doubt that there was.

Finally, on Sol 21, August 10, 1976, the scientists of the Viking Biology Team announced the results: either Mars had some forms of microorganisms in the soil around the lander site or the soil had a chemistry that was not easy to duplicate in Earth laboratories and a soil chemistry, moreover, unknown here. "Well," Harold Klein of the Viking Biology Team was asked, "does that mean there's life on Mars or not?" The answer he gave could have been the response to a question about Mars posed several hundred years ago: "Well, the question is whether Mars is talking with a forked tongue or giving us the straight dope." In other words, maybe so, maybe not.

The data was not quite as ambiguous as the Biology Team made it. What the scientists were trying to say was fairly simple. Had the results produced from the soil been encountered on Earth, we would certainly have concluded they represented biology. But since they were encountered on Mars, there was the possibility, as someone put it, "of the *Martian* explanation." Mars *may* have some bizarre soil chemistry that acts a little like biology. Results from *Viking Lander 2* were much the same; it landed on Mars September 3.

Finding life was only the most publicized scientific effort of the mission. The orbiters above photographed and mapped the planet to a degree unheard of in the dark days of *Mariner 4* and its miserably bleak photos. The two landers charted the weather and climate at the surface.

The original plan had been a ninety-day mission, although everyone hoped for much longer. They got their wish. *Orbiter 2* died on July 25, 1978, with the fuel for its attitude control system exhausted; it could no longer keep its solar panels on the sun line. *Orbiter 1* functioned until August two years later when it too ran out of attitude control gas. The two satellites between them had transmitted 57,000 photos of Mars to Earth. They mapped 97 percent of the surface, showing objects as small as a thousand feet in diameter.

The landers had even greater longevity because they had been designed to function even if the companion orbiter died, but by April 1980, the pictures and weather data from *Lander 2* were gibberish. It had taken 1,800 photos in its far northern location. It sat on Mars, cold, unpowered, and silent, surrounded by the thin layer of water frost that covered the ground around it in winter.

Viking Lander 1 seemed to be headed toward a place in history as the "ancient one" of Mars. Six years after the landing, it had photographed every square inch of the terrain within its vision several times over. It had reported that the weather varied only a little except for occasional dust storms, some of which were planet-wide. The lander's potential for long-range Martian reporting was good. Placed in what the scientists called the "eternal mode," it could operate almost unattended until 1994, sending a weekly weather report and pictures to Earth.

On November 5, 1982, after 2,238 Martian days on the surface, the last lander's computer dutifully turned on its cameras for a picture of the surroundings. The photo was radioed to Earth on November 13, right on schedule. It showed dark-colored ground and vague shadows, indicating that a dust storm noticed several weeks earlier was still under way. The photo was part of a long-range series to monitor the scene under identical lighting conditions. Scientists were looking for changes in the ground and in the atmosphere, changes that were slow on Mars compared to Earth, and the lander was doing a version of time-lapse photography.

The next contact was scheduled for November 19. It never came. There was no immediate indication of what might be wrong; the lander had been functioning perfectly. JPL engineers tried several times to transmit new commands to the spacecraft computer; they failed. It had been previously programmed to begin sending on its own if it did not hear from Earth for a period of nine consecutive weeks. The engineers and scientists waited until May 5, 1983. There was no call home from *Viking*. Finally, they tried a new trick: they commanded the lander's computer to reorient the communications antenna, just in case it had lost its "lock" on Earth. There was silence.

At last, on May 21, 1983, JPL officials reported to NASA that it appeared to be permanently out of communication with Earth; they recommended that the Viking mission be terminated after almost seven years. The final agony came in Building 230 at JPL. The darkened monitor that was attached to the computers and to the receiving equipment on Earth showed nothing but a flat, white line, as flat as that of a hospital patient past the point of no return. The first major investigation of Mars from the surface was finished.

There is one footnote: *Viking Lander 1* was renamed the Thomas A. Mutch Memorial Station after the NASA chief scientist who had been the leader of the scientific team responsible for the photographic efforts. Dr. Mutch had been killed in a tragic mountain climbing accident in the Himalayas. In addition to renaming the lander in his honor, NASA created a stainless steel plaque that read: "Dedicated to the memory of Tim Mutch, whose imagination, verve and resolve contributed greatly to the exploration of the solar system."

The odd part of the story is that the plaque contained an additional line. That line was marked "Emplaced" and left blank for another date. NASA administrator Robert Frosch announced that he was charging a future administrator with responsibility for attaching the plaque to the lander that lies at 22.48°N latitude, 47.97° longitude, on the dusty slopes of Chryse Planitia.

In adding that line, Robert Frosch was not simply being sentimental. He knew we'd go to Mars in person someday. He knew we would one day stand on that plain and stare at the weathered little lander under the pink Martian sky. And that is what we will do.

2 THE RHYME OF THE LAST MARINER

THE RESULT of the Viking exploration is that Mars is not unlike Earth in *gross* planetary terms. This conclusion is extremely important if we want to go there and live someday. Mars is arid, cold, and windy. The atmosphere is presently inimical to humans. The sky is a disgusting pink.

But except for the bad air, the high ultraviolet bath, the pastel sky, and the dim sunlight, Mars is not much worse than Antarctica on a bad day. Actually, 99 percent of us couldn't distinguish a Martian rock from one of our own. The sand is almost the same as the grainy stuff that gets in our tennis shoes. There are mountains, jagged peaks, and deep arroyos. It is not the most inviting place imaginable, but it is not the dead and sterile lunar surface, either.

There seem to be vast quantities of water in the polar caps, estimated at various times at between three feet and half a mile thick. There is unquestionably a large amount of water bound up in the soil as permafrost, which is not unlike that found in Alaska or Siberia, except that the biological count appears to be nonexistent.

There is water vapor in the atmosphere, less than on Earth, but significant. And there is "snow," or at least a thin layer of water-frost on the deserts and mountain chains at times; some even came overnight onto the *Viking* landers in winter. The scene was obediently photographed by the idiot machines; the frost made it look remarkably Earth-like.

There is sufficient sunlight to cause noonday summer temperatures to come within a range we humans call comfortable. When Mars is closest to the sun, it receives about half as much sunlight as we do; when most distant, slightly better than a third. Summer on Mars, in other words, is a little like northern Canada in winter in the amount of light it receives.

The dust storms are often bad: but then, winds are often bad here, and we survive them. People who live in Florida do so without paying much attention to hurricanes except during the season. People live in the shadow of volcanoes, perch on the rims of mountains; they build houses on geologic faults and in mudslide areas. The human being is a daring creature when it comes to battling nature. That element of our psyche will be useful.

While there *is* some water vapor, plenty of ice, and vast regions of permafrost, there are no supplies of liquid water on Mars today. That is, with the possible exception of the area around Solis Lacus. The Lake of the Sun, Solis Lacus, is a kind of mythical place like the Tropic of Cancer is on Earth. It is the spot on the planet closest to the sun when Mars is at the most sunward point in its orbit. It is probably a Martian Death Valley, with blazing temperatures in summer — perhaps as high as 75°F, although the nights there are frigid.

Solis Lacus may be a real oasis on Mars, a kind of Martian bog. The scientific evidence — Earth-based radar, *Mariner* and *Viking* — suggests that fairly large amounts of water vapor are periodically "outgassed" (outgassing is what Mount St. Helens did: a sort of planetary burp) from the area. This means that the region may have a body of water buried in the subsurface soil that occasionally, on very hot days, comes to the surface and evaporates into the air. The amount is greater during spring and summer, less during fall and winter. There may even be times when Solis Lacus has liquid water for short periods.

The question of whether there is water on Mars is a crucial one. If the water were reasonably available to an exploration party, planning a mission to Mars would be much easier. Water is a basic ingredient of human life, of course, and it is extremely heavy. Mars is much farther than the moon and a rocket would need an immense amount of fuel to carry enough water for the trip, exploration of the planet, and the return voyage. Tanks for transporting the water would also add to the weight of the rocket.

Unlike food supplies, water cannot be "freeze-dried." The old saw about packages of freeze-dried water — "Just add water as required" — falls somewhere between the search for the Holy Grail and perpetual motion machines.

Water can be chemically synthesized aboard a spaceship; the space shuttle uses a system that synthesizes water as a by-product of rocket fuel, but that requires extra fuel, and the extra fuel requires a larger rocket. It's a circular problem. Martian supplies of water break the circle and would be particularly handy for any colonization project.

Carrying oxygen to Mars is also a costly proposition. And importing vast quantities of it for a permanent base or a Martian colony is impossibly expensive. But what if oxygen were available on the planet? One of the most elementary experiments in every school science class is the electrolysis of water. The result of electrolysis of H_2O is hydrogen and oxygen. Hydrogen, in liquid form, is the most common rocket fuel, used with liquid oxygen. Oxygen as a gas is what we need in air to sustain life.

While oxygen can be provided from Martian water supplies, it is much more profitably obtained from the soil. The soil compounds of Mars, because of their rather odd chemistry (they are super-oxidized), are extremely useful: when heated, they produce oxygen.

Carting food to Mars, even the freeze-dried variety, is also expensive in terms of fuel costs. While the present soil of the Red Planet is deadly to Earth's plants (some of the acids are dangerous), with the right imported soil cultures an exploration party could be relatively self-sustaining. Later, when a permanent base was established, it might be possible to grow plants directly in modified Martian dirt.

Let us return to the honeycomb aluminum ladder that our unprotected human descended, and let's watch the human descend again. Death is not the only possible outcome, after all.

The human presses the switch in the airlock and the door swings open to reveal the Martian surface. The human is wearing a battery-heated, reflective suit impervious to the ultraviolet rays of the Martian sun. The suit is grafted onto protective boots that will make the first ribbed footprint on Mars. There is a clear plastic helmet, mildly tinted against the sun's glare. It is locked to the environment suit by a quick twist to the right. Strapped around the back of the suit is a body pack made up of the batteries and oxygen flasks.

The human takes a first step down the ladder. Three rungs from the bottom, the human leaps into the thin Martian air and lands 20 feet away; it's not much harder than scuba diving in cold water, and the equipment is somewhat similar.

Our traveler and the other humans who descend the ladder haul a packing crate from the spaceship's cargo hold. In the 40 percent gravity, the crate weighs much less than on Earth, but it is just as bulky and hard to move. Inside the crate is a small solar-powered heating plant. It is erected at Solis Lacus. The solar panels are unfolded and a load of soil is placed in a hopper. At the other end of the compact machine, pure oxygen is pumped into flasks for use with the environment suits.

Within a few days the explorers have erected a small greenhouse, using a solar-powered tractor (or nuclear-powered, if our technology is advanced enough) to dig the foundation. The dug-out soil and dirt go into the oxygen generator. The greenhouse is a giant plastic dome, much like those built for swimming pools in the Colorado mountains for year-round use. The dome, partially buried in dirt, is heated by a solar-electric generator and filled with an atmosphere rich in carbon dioxide, of which Mars has more than enough since that is the primary constituent of the air.

A few days later, the dome is filled with collapsible benches and trays. The trays contain soil cultures from Earth with perhaps a mixture of carefully selected Martian compounds that will not inhibit the growth of Earth's plants. The trays are moistened by water taken from the permafrost of the landing site. The seeds are from Earth. It will not be long before a number of carefully chosen vegetables could be available to the members of the expedition.

Roses (red ones, of course) on the table of the first expedition to Mars? It is possible but perhaps not likely. The "farm" of the first expedition will be entirely experimental, providing scientific information that will be useful to later explorations and a permanent base. The first explorers' food will come from Earth — freeze-dried, much to their boredom.

In a week the explorers have erected an additional buried structure, another plastic Quonset dome, and, suddenly, we are Martians. It isn't Miami, but it is home on Mars, and not completely uncomfortable. Using this sort of technology — and other, more advanced technologies being developed — we could soon land and survive for months, perhaps years.

Fine, we can live on Mars. It is a relatively Earth-like world, and with some wonderful gadgets we could get along quite nicely for an extended stay. But could we get all these: electrolysis plants, nuclear-powered tractors, astronauts, domes, and plant trays all the way from Earth to the fourth planet and get some of it back safely? After all, *Viking* weighed hardly a ton. We are speaking of a manned mission that could involve hundreds of tons. So can we do it?

You have to be cautious in asking this sort of question. Any good astronautical engineer will tell you it is *possible* to attach a rocket engine to a horizontal VW bus, tilt the bus slightly, and make it fly up — to the moon or even to Mars, if that is where you want to go. The engineer may not mention that the raw power required to do this bit of propulsion legerdemain is staggering. That's why rockets are generally slim and pointed and don't look like VW buses. It keeps the power requirements somewhere within the realm of the engines we have available.

The short answer is this: yes, we can fly a manned mission to the Red Planet. We *might* be able to do it by the turn of the century or shortly thereafter. It *could* cost less than the landing on the moon. There is very little in the technology required that is not currently under development or cannot be produced from already existing designs.

Going to Mars is a relatively recent dream. Until late in the nineteenth century, the moon was the most popular destination. And the concept of even the most rudimentary spaceship was not possible until the Industrial Revolution gave us the technology of great machines and powerful engines.

What fired the imagination about our favorite planet was the long and continuous study of it through the last hundred years. It was the planet most easily studied — apart from Venus, about which nothing could be learned because its surface was always hidden in clouds. Through years of lonely nights at telescopes, astronomers pieced together what they thought was an accurate picture of the fourth planet from the sun. It was a picture that would still haunt us right into the age of spacecraft.

In the eyes of the public and many scientists, the most significant event in planetary astronomy before the use of spacecraft came in 1877. Giovanni Virginio Schiaparelli, an Italian astronomer, reported that there was a strange and intriguing thing about the light patches on Mars which had previously been classified as land. In the light land areas, he said, were thin but clearly visible lines that connected the dark and light regions of the planet with the polar caps. (This "discovery" was not completely new: actually, there are lines on maps of Mars dating back to 1840.) Schiaparelli thought the lines were at a lower level than the surrounding landscape, so he called them "grooves." The Italian word for grooves is *canali*. When his reports were translated into English, we got the "canals of Mars," which implied inhabitants.

Words stick quickly in the mind. When William Whewell used the word "scientist" for the first time, it stuck. A drunken poet told Galileo that "optic tube" would simply not do for something so grand, and so we have "telescope."

Jean Paul Sartre never used the word "existentialism" — it was coined by a Paris journalist. Thankfully, the word once proposed to describe the inhabitants of Mars — *Marticoli* — did *not* stick; instead we have Martians.

The public was convinced by books such as *The Planet Mars* by Camille Flammarion, who argued skillfully in favor of a canal-building civilization of superior beings. The public fancy became so great that giant mirrors were proposed for signaling the sister race on the Red World. The press — sensationalist and conservative alike — had a wonderful time. The *New York Tribune* reported a claim that certain dark markings on Mars spelled out the word "Almighty" in Hebrew. No one asked how a knowledge of Hebrew, ancient or modern, could have been acquired.

Onto this grand stage of public excitement stepped three men whose lives and thoughts would forever be inextricably intertwined with the fate of the planet. Percival Lowell was one. A young English writer of what would later be called science fiction, H. G. Wells, was another; and Edgar Rice Burroughs, who wrote the Tarzan books, was the third. They put the word "Martian" into nearly every language on earth. They gave us the grand dream of the Red Planet.

Of the three, the most influential and probably the most controversial was Percival Lowell. He was born in Boston on March 13, 1855, a product of two of America's greatest and wealthiest families, the Lawrences and the Lowells. His sister was Amy Lowell, a first-rate poet; his brother, Abbott Lawrence Lowell, became the president of Harvard.

The discoveries of Schiaparelli profoundly influenced Percival Lowell, and he founded an observatory in Flagstaff, Arizona, on a small knoll called Mars Hill. At that time Arizona was still a territory — statehood was almost a decade away. The mile-high desert air was ideal for observing the planets: it was clear and steady, with no city lights to interfere. The observatory was opened in 1894.

Lowell — first single-handedly, then later with assistants — began a systematic study of Mars. So intense was his effort that even in his own lifetime he was acknowledged as having made one of the most remarkable investigations in the history of astronomy. He escalated what had been a rather mild debate about the canals into a public sensation. So great was his impact that the *Wall Street Journal* in its roundup of 1907 listed proof of "conscious, intelligent life" on Mars as one of the highlights of the year.

Lowell's first popular book — a best seller, published by Houghton Mifflin of Boston — came out in 1895. In it he set out a confident picture of the Martian world: it was an ancient planet, slowly losing its life-giving water. Lowell calculated that Earth had 189,000 times more water than Mars. The canals, Lowell suggested, were vast irrigation projects built by the inhabitants. The great artificial channels fed water to the fading deserts. The lumpy places where the channels intersected — the oases — were enormous pumping stations. What we saw in our telescopes was the wide band of vegetation bordering the canals. The channels themselves were too narrow to be seen from Earth with telescopes.

He thought the lack of water would have caused any life to move from the sea to the land at a much earlier date than did Earth's creatures; hence life on Mars was much older. He also thought that because they had to cope with their dying planet, the Martians had evolved into a highly intelligent race more quickly than we did. (Lowell was at heart an unrepentant Darwinian.) Lowell's Martians were peaceful and contemplative, devoid of the complications and pestilences Earth's people often infected themselves with. His Martians were a kind of gentle Greek chorus compared to the strident voices of mankind. He fed material to every science fiction writer in Christendom up to and including Ray Bradbury, who wrote that most haunting and beautiful eulogy of the dying race, *The Martian Chronicles,* more than half a century later.

When the public was through reading Lowell's three books, the last of which was published in 1908, there was little question that Mars was inhabited. We believed in the Martians as truly as we believed in the progress of technology, the watchword of the era. C. E. Housden in 1914 went into great detail in describing Lowell's Mars. He specified the nature and arrangement of the great pumping stations that fed water from the poles to the equator via the canals. The water-bearing pipes, he figured, were 6 feet in diameter. All this technical data was surprisingly real — if one assumed that Lowell's description was real.

Adding to the public consciousness was the tremendous impact of H. G. Wells's *War of the Worlds,* published as a book in 1898. Herbert George Wells drew heavily on Lowell for his planet, but of the Martians he took a different view. In sheer relentlessness, the fictional attack of the Martians on Earth's inhabitants has had few equals. It is perhaps because H. G. Wells studied biology under T. H. Huxley that his solution to the invasion was a biological one.

The third member of the triumvirate, Edgar Rice Burroughs, wrote his first story about Mars when the theories of Percival Lowell were still prominent. It was 1911. The Lowellian planet, the world of a dying Martian race, dead sea bottoms, and ancient wisdom, was translated by the young author into the imaginary planet Barsoom. In *Princess of Mars* and a dozen other books, Burroughs featured John Carter, warlord of Barsoom. In popularity and sales, they were overshadowed only by his Tarzan series. Thousands of young men — some of whom later became rocket engineers — fell in love with the fictional Martian princess Dejah Thoris.

H. G. Wells died after World War II, having lived to see photos of the first giant liquid-fuel rockets invented by the Germans — rockets whose mechanical descendants could land on his terrifying planet. Burroughs had a longer span: he died only fifteen years before a spacecraft showed a world far lonelier than his wonderful Barsoom. Having influenced not only Wells and Burroughs but his entire generation, Lowell died in 1916, before the increasingly technological world of science began systematically destroying his vision of Mars.

The public was not, in any event, very interested in the speculations and experiments of twentieth-century scientists. Mars had been fixed in the mind as an inhabited world with artificially constructed canals, and there it stayed. The fact that Percival Lowell had drawn lines resembling canals on both Venus and Mercury went unremarked. So, too, did the temperature figures obtained in 1926 ($-150°$ F at night) and the fact that no oxygen could be found in the atmosphere in 1930.

The captivating fantasy of a dying Martian race fueled immense speculation about going there to meet them. While it might be argued that Lowell, Wells, and Burroughs created the destination, it was another writer who most deeply influenced the discussions about transportation. His name was Kurd Lasswitz; he was born in Breslau, Silesia, in 1848. The book was called *Auf Zwei Planeten* (On Two Planets). It was published in October 1897 and appeared in ten languages — none of them English.

In his novel Lasswitz subscribed to the Lowellian view. The Martian race is running out of water. To avoid growing things that use water, they eat synthetic foods. Lasswitz provided his fictional Martians with rolling roadways, antedating by more than half a century science fiction author Robert A. Heinlein's famous "rolling roads." He also had the inhabitants of Mars deeply engaged in building space stations over Earth to help them transfer materials and "colonists" to Earth's surface. They also have space stations over their own planet. It helps in interplanetary travel, the Martians explain.

The most interesting parts are the solutions to space travel. Kurd Lasswitz was a mathematician and scientist first, a novelist second. His spaceship voyage is scientifically accurate, although the propulsion is a touch creaky: it's a variation of antigravity. The novel contains a detailed explanation of how to travel between worlds: a spaceship leaves the immediate gravitational field of one planet and then assumes an orbit which will eventually intersect that of the destination planet. Spaceships do not fly in a straight line.

According to Lasswitz, the time of transit can be calculated; it depends on the time of departure and on whether one intersects the orbit of the other world at the right time. Wells, Verne, and other imaginative writers were ignorant of this fundamental issue of real space travel. Lasswitz also invented the retrofire sequence, which allows any spaceship to change orbits, and the midcourse correction, so much a part of the jargon of the space age, although he never called them by those names.

The importance of Lasswitz's *Auf Zwei Planeten* is this: while the earliest rocket pioneers were heavily influenced by Wells and Verne, the ones who came later were deeply in debt to Lasswitz. His legacy was talk of space stations and of cargo-carrying spaceships to serve them and the planet below. It was the Lasswitz novel that prompted the first *serious* discussions of how to get from one world to another. Among the more intent readers of the book were Walter Hohmann and a very young German named Wernher von Braun. Hohmann gave us the first engineering figures for reaching another world using the least rocket fuel. The mathematical method is still called the Hohmann Orbit and could be used by the first manned spaceship to visit Mars and by our grandchildren when they settle on the planet. Dr. von Braun gave us quite a bit more, almost everything, in fact.

One quiet evening in 1947 at Fort Bliss, where Dr. von Braun was working for the U.S. Army, he picked up a slide rule and worked out a dream. Using only the advances inherent in the wartime V-2 rocket, which he had helped develop, he designed a complete mission to Mars. After all the pulp stories, after the planet had been reached in fiction by everything from teleportation to antigravity, here was the first real scientific approach to getting there.

In *Das Mars Projekt* von Braun produced a plan involving ten spaceships with a total crew of seventy. The ships would be assembled in a two-hour orbit high above the earth. The materials for building the ships would be taken to the construction site by giant three-stage ferry rockets. As the propellant, structures, supplies, and crew arrived, a work force would assemble the expedition.

The ten ships would depart the construction site and assume a long elliptical orbit around the sun; the path would touch the orbit of Mars eventually, and when the ships got to that point, they would fire their rocket engines and take up station around the Red Planet. This is exactly what *Mariner 9* did, nearly a quarter of a century after von Braun proposed his mission. But von Braun was writing about a manned mission, not an unmanned probe like *Mariner 9,* so three of his ten ships were equipped with "landing boats." Leaving the motherships, the surface party would descend to the Martian surface. The boats were equipped for a glider landing and a vertical rocket takeoff — they were smaller than, but otherwise not unlike, the space shuttle that von Braun was later to help design.

For a return to Earth, two boats would bring the surface party to the orbit around Mars, where they would board seven of the original ships, leaving the landing-boat motherships to circle the ochre globe of the fourth planet forever. The seven ships would depart and enter a long elliptical orbit around the sun until they came near Earth's gravitation field. They would then fire their rockets and change their orbit to one around the home planet. The crews would be taken down to the planet aboard one of the upper stages of a ferry ship.

The entire mission would have taken about 520 days flight time, plus time spent in Mars orbit and time on the surface. It was all worked out: rocket engine thrust, heat shields, propellants, interplanetary radio, tracking, midcourse corrections, the possibility of cosmic-ray damage to the crew, the long voyage without gravity, the danger to the spaceship of bombardment by meteors. It was a complex and detailed analysis of how to get to Mars, survive there for an exploration, and return safely to Earth.

Time, however, and the march of technology have caught up with some of von Braun's ideas, as he acknowledged in a new introduction to *Das Mars Projekt* in 1962. By then, rocket engines that were twice as powerful as those he envisioned were already available and powering the giant Saturn V rockets that would land man on the moon seven years later. Much of what von Braun had made educated guesses about in 1947–48 was known by 1962; he thought then that a landing on Mars was as close as fifteen or twenty years away. It finally dawned on the world that von Braun and others like him were not crazed, semilucid visionaries or pulp science fiction writers. In a lifetime devoted to the single issue of conquering interplanetary space, he had amassed twenty-five honorary PhDs. He was a member of thirty-one professional societies, ranging from the ultra-exclusive Explorer's Club to the British Interplanetary Society. He was extremely fond of wildflowers.

Following the landing on the moon, it was assumed by many people in NASA that Mars was the next target. In fact, many of them thought of the lunar landing as a warm-up, as indeed it always had been considered, from the worthy German pioneers of rocketry on down. The only problem with this idea was that it ignored politics. When most people think of technology they think of it as a single entity, which it is not. Technology is a blend of science, engineering, and application. The scientist tells you what is possible, not necessarily if it is practical. Making it practical is the job of the engineer. But even practical ideas may not be put into motion — application is a political issue, given that large technological efforts such as landing people on the moon or on Mars are in the province of governmental organizations, subject to the infighting that is part of all political institutions.

By the time Buzz Aldrin and Neil Armstrong stepped onto the moon, NASA had finalized a standard Mars scenario. It involved one Saturn V moon rocket that would launch a manned spacecraft. Five more Saturn V's would carry a nuclear propulsion module up into space. Launching the assembly crew and the mission personnel would be done by smaller Saturn 1B boosters. The five big rockets represented a massive launch weight, nearly half the weight of the entire mission.

The plans by von Braun and NASA were all very expensive, though quite possible, either in 1948, 1962, or 1971. There was nothing there that was *wrong;* the plans were just extremely complex. Part of that complexity is a legacy of Kurd Lasswitz: he said the best way to visit Mars was to have orbiting space stations over both Mars and Earth and use freight-carrying "cars" to take materials and crew back and forth. The actual interplanetary journey would be done by a ship that was never designed to land anywhere.

It is a little unclear exactly how Lasswitz hit upon his answer, but he was right. Even with the most advanced of rocket engines burning liquid fuels, no rocketship can leave the surface of Earth and travel to the surface of Mars. The reason is simple: it takes an enormous amount of fuel to escape Earth's gravity. Even the gigantic Saturn V lunar rocket could deliver only a hundred tons or so to the moon. A rocket would have to be many times as large to have any chance of reaching Mars with the same hundred tons. Since this has been known since the earliest days of rockets (the problem has been called "the gravity well"), it has always been assumed that any interplanetary spaceship outside of fiction would be built and launched from orbit. There's an old saying in the space business: "Once you're in Earth orbit, you're halfway to anywhere." It's true.

There was always a second part of the equation: was there any point in going there? By the time NASA did its study in 1969, there was serious doubt in the minds of many scientists whether Mars was worth visiting by manned spacecraft. While the public might still believe in Martians — Hollywood saw to that quite well — the astronomers, chemists, physicists, and biologists who were still studying the Red Planet felt differently.

Did it not affect us all, this casual scientific debasement of our dreamworld Martians tending their dying planet? In all the long time we had discussed inhabited worlds, we had hopes for something else out there, some haven, some other Earth. We wanted not to be alone in this cold and infinite universe. That was the reassurance we wanted.

We didn't cease dreaming; we just changed the emphasis.

If Mars wasn't inhabited, then *we* could be the inhabitants. We also began to wonder, pointedly, whether the best way to get there was by using gigantic NASA missions costing billions. If we wanted to live on Mars one day, not just go there and stare at the tiny blue star in the pastel-pink Martian sky that is Earth, we might try something other than a singleminded one-shot exploration mission. What we wanted was a way to sustain the exploration of Mars, establish a permanent base, and above all do it cheaply.

3 GATEWAY TO MARS

THE SPACE STATION will lie in an orbit several hundred miles above Earth. The black around it will be as black as Margaret Thatcher's Daimler heading for Whitehall, as black as the proverbial lining of the Earl of Hell's waistcoat. It will be cut by hard, bright star points everywhere, diamond dust sprinkled except where broken by the sphere of the pale, off-white moon. From up there, the moon no longer looks like a coin suspended in the sky; it is a ball, a globe, a distinct and lifeless world. Below will be the greens, browns, and blues of Earth, perhaps with the continent of Africa just visible under a clump of dirty weather.

In orbit, the terms "above" and "below" have no meaning, but it is sometimes still convenient to think in the old expressions of home, several hundred miles away. Our thousands of generations on Earth go back that far, are buried that deep; our thrust into this new environment of living in space is that new.

The constellations will be deeply etched in the endless black, the imaginary shapes unbelievably clear. The Great Andromeda Galaxy will hang as a misty oval, a reminder that we are but a small part of only one galaxy which we call the Milky Way. Perhaps our Milky Way is not even a very important galaxy. Perhaps it's a statistical backwater, a galactic small town complete with the equivalents of a high school marching band and picnics in the parks, filled with people whose accomplishments are important to themselves but are only a pale flicker of the sophistication elsewhere. Perhaps men and women not unlike us in one of those distant galaxies dreamed, designed, and set out on a journey to their nearby planets long before one of our own ancient ancestors amused himself on the shore of some primordial sea by scratching designs in the sand.

Seen from the space station, the Magellanic Clouds would stretch across part of the blackness like patches of dusty cotton pasted onto a velvet backdrop. If we set out for those clouds in our Mars spaceship, the sun would go nova before we ever arrived; a billion years might pass. The clouds are beyond the Milky Way, our own great lens-shaped wheel a hundred thousand light years in diameter and filled with two hundred billion stars, only one of which is the yellow G-type star we call the sun.

The ship that will go to Mars will lie in orbit, separated by fifty miles from the space station. It will be an awkward, ungainly ship; it will never enter an atmosphere and there is no reason for it to be aerodynamic. Plying the space between the ship and the station like lighters servicing a tramp steamer will be space tugs, little cargo barges carrying supplies.

We will be ready to start the great dream, to fly over the dark miles that separate us from the Red Planet and land there to stand beneath the strange pink sky and the jumbled rocks left over from some cosmic upheaval. We will have waited a long time for the chance: nearly four hundred years since a philosopher said there were other worlds and that they might be inhabited; nearly seventy-five since the first successful flight of a liquid-fuel rocket; more than a quarter of a century since we took our first footsteps on the moon. We may feel the ghostly presence of the dreamers: Schiaparelli, Lowell, Wells, Burroughs, Lasswitz, Goddard, von Braun . . . They will be there.

In the distance, but still the closest object to the ship, will be the great gleaming spirals and modules, the disk antennas and spidery instrument platforms of the space station. A third-generation space shuttle will arrive and dock at the main supply module. Among the thousands of tons of supplies it will be unloading are a long-awaited tape of *The Maltese Falcon*, a box of genuine Havana cigars, and fifteen decks of plastic playing cards, one a deck of Tarots.

The crew of Lego City, as the station may be called, will be the last remaining human contact: the men and women of the orbiting laboratory will be the final few to say farewell and Godspeed to a small segment of their kind who are about to set forth on a voyage of exploration into the unknown. The "aether," as a nineteenth-century man would have called it (to us, UHF radio frequencies), will be full of NASA-talk. "You are ten seconds from ignition," a disembodied voice from Combined Space Command will rumble. "Ten, nine, eight, seven, six . . ."

The countdown was the invention of Fritz Lang for *By Rocket to the Moon*, a movie made in 1929. The script was by Thea von Harbou, otherwise known as Mrs. Fritz Lang — no one missed a Fritz Lang film. Looking around for a gimmick for the film, Lang had hit upon the countdown. His intuition said it contained a vital dramatic element, and it does.

"Five, four, three, two, one, ignition. *Mars-1*, we show ignition." "Roger, CSC, we have ignition." The engines of *Mars-1* will ignite and the ship will slowly leave its orbit around Earth. Telescopic TV cameras aboard the space station will record the event for the special bulletins, the early and late news broadcasts. The reporter aboard the space station will give a studied commentary with frequent references to scientific experts both at the station and in home studios across the world.

Mars-1 will move away from the arrays of communications antennas, the astronomical observatory and space telescope. It will drift away from weather satellites, the navigation orbiters, the great military and commercial TV relays and transponders in high geosynchronous paths at 22,500 miles, and slowly disappear. It will be leaving all that is known and all that has come about since the space age began. It will move into the darkness toward that red point of light which it will not reach for months.

Can the exploration of Mars really begin this way in less than two decades? The answer is a qualified yes and a slight wagging of the finger to discourage too much optimism. Even designing a Mars mission that uses almost exclusively off-the-shelf stuff — and that is at least *possible* now — must take into account the inevitable time lag in actually getting it all together. The space shuttle was almost ten years in development, for example, before it went into space for the first time. The lunar landings also took nearly a decade from conceptualization to success. A Mars mission depends primarily on the successful completion of several projects in space that are scheduled to occur over the next ten years. If those other projects are delayed for any reason — economic, political, or technological — then any Martian voyage of discovery will also be delayed.

The first step toward Mars is some method, preferably practical and cheap, of sending materials into orbit, where they can be used to build a spaceship. One of von Braun's later proposals, written long after the age of space began, envisioned *four hundred* trips with an expendable ferry vehicle. The cost of four hundred expendable rockets to make the shipments would prohibit any expedition within the normal budget and politics of any nation of Earth. Fortunately, we have a reusable cargo ferry called the space shuttle which will eliminate the need for four hundred ferry rockets.

The space shuttle needs little introduction, but perhaps a bit of history would help. It is a direct descendant of German rocket engineering and planning near the end of World War II. Some of the on-paper rockets the Germans were toying with had winged third stages and intermediate stages developed from the V-2. Wernher von Braun later popularized the design of a winged, reusable, third-stage rocketship in the 1950s. But despite the earlier design work, all of the launches of the United States and other countries until the space shuttle came along were by expendable rocket.

Expendable rockets are just that: expendable. The first and second stages fall to earth, land in the ocean, and are lost forever. Sometimes the third stage achieves orbit, sometimes not, depending on the design and the orbit it is trying to send a payload into. Such rockets do not deliver a fantastic weight into space, considering that the majority of the cost is lost in the bottom of the ocean or when the debris burns up in the atmosphere about a decade later. Expendable means expensive.

The space shuttle, on the other hand, is a very economical

vehicle. The solid rocket boosters that start it off the launch pad are recoverable and are used again after each flight. The main spaceship, the orbiter, lands back on earth after each trip and can make hundreds of voyages. The only part that is not very cost-effective is the main fuel tank, called the external tank; currently that falls into the Indian Ocean and is wasted. The cost saving over using expendable rockets is enormous; it makes space exploration practical for the first time. Almost anything can be built in space if enough materials and supplies are hauled up by the shuttle, including a spaceship to Mars.

The current space shuttle will be modified in the next few years with a "liquid boost module," an additional fuel supply. It will be able to carry heavier loads. There are also improved engines on the design boards that will increase the cargo-carrying capacity. Plans for second- and third-generation space shuttles envision gigantic unmanned boxcar modules that can be attached and sent into orbit. The importance of the present shuttle and these future freight-carrying boxcars is that they will be paid for and available during the development of the shuttle program — and for building the American space station. Only *four or five* launchings of these super cargo carriers might be needed to build a Mars ship in a parking orbit several hundred miles up. If the shuttle is the first step toward Mars, the space station is the second.

The idea of a space station is not new. Sir Isaac Newton predicted artificial "moons" several centuries ago, and manned artificial stations have been discussed by many of the pioneers of space travel. The ever-visionary von Braun popularized the idea of a wheel-shaped space station in the 1950s, at the same time that he described a ferry or shuttle ship to supply it. Before the Apollo program had successfully landed men on the moon, engineers and scientists were convinced that a space station was the next step, that is, after they designed a ferry vehicle. In actuality the order was slightly reversed, since a giant, expendable Saturn V moonship launched the first American space station, *Skylab,* on May 14, 1973, years before the ferry-shuttle became operational.

Skylab was pivotal to any idea of long space flights to the planets. Flights aboard the orbiting laboratory proved that people could live and work effectively in a weightless environment; they proved that for life to proceed normally it was not necessary to provide an artificial gravity by spinning the station. It was our first house in space, and it proved that we could exist there for at least twelve weeks and return to Earth without ill effects.

But *Skylab* was not a *real* space station because it could not easily maintain its orbit (it eventually came too close to our atmosphere and was burned up), and its design did not lend itself to expansion. What the aerospace engineers had in mind was something more permanent. Unfortunately, the political bickering over the tremendous expense of space exploration got in the way. It has taken more than a decade for another space station to be funded.

But when it is in place sometime in the early 1990s, we will have the way to Mars open. The shuttle can send materials into orbit so that an expedition ship can be built and the space station can be the construction base, complete with workers to do the job. The only thing left is to design a spaceship.

A hundred NASA engineers have designed manned vehicles that could reach the Red Planet, including the old 1969 concept that relied so heavily on Apollo lunar hardware. An equal number of independent scientists, engineers, and space enthusiasts have already addressed the problem, especially in several major gatherings called "The Case for Mars" conferences, the last of which was held in 1984.

One of the more innovative ideas to save money going to Mars involved the use of space shuttle external tanks. The tanks do not now achieve orbit, but a slight extra push would put them there. They could then be joined into an ugly but perfectly serviceable spaceship by astronaut construction workers. Each tank is about the size of a 747 without wings.

Using shuttle tanks as a "Mars Transit Vehicle" is an interesting solution because the first exploration of Mars could be done for less than a *tenth* the cost of the lunar exploration, according to some estimates. Unfortunately, quite a few engineers studying the problem have concluded that it would eventually be cheaper to design a spaceship from scratch.

So perhaps we need a *real* spaceship for our Martian expedition. There is nothing in designing one that is beyond our present capabilities. It would be an expensive and complex effort, but so was the space shuttle. What seems most likely, despite fantastically innovative ideas that have been put forth (such as ion-powered spaceships, modular ships in trefoil configuration that have artificial gravity), is that the first manned rocket to Mars will be a traditionally constructed and traditionally powered vehicle.

A gas-core nuclear rocket could provide plenty of thrust for a spacecraft the same weight as a loaded 737 — and for a couple of million pounds of extra fuel. A ship like that, complete with five astronauts and a life-support system, could go from Earth to Mars and back in sixty days. The problem is that we don't yet know how to build one and some physicists working on gas-core nuclear designs don't think we ever will.

Ion rockets are presently under development. They have advantages for colonizing Mars, since the argon of the Martian atmosphere can be used as a fuel for returning to Earth. However, they are not sufficiently developed to be useful now, and in any event they would have to be supplemented by regular engines much of the time.

So the first Mars expedition will probably go up the hard way, the old von Braun way: chemical fuels to the space station by a large number of shuttle loads, and then construction materials sent into orbit the same way. The ship will be developed from space station technology and from shuttle engineering.

The cost will be high, perhaps as much as $35 billion if too much new research and development work has to be done, perhaps as low as $10 billion if enough already-paid-for technology is applied. At best, it will be much cheaper than the *Apollo* landing on the moon; at worst, less expensive than the infamous "Star Wars" defense system.

Some of the money may come from private sources. Mars has always been a success when it comes to fund raising. The Viking fund, a grassroots effort led by private individuals to help finance the exploration of the planet, brought in more than $60,000 in the first year. While this sort of thing is hardly enough to fund a mission, it is worth pointing out that Americans spend more than $100 million a year on science fiction and space fantasy. If that same amount could somehow be encouraged away from the same people for a real space adventure, that source might provide one-thirtieth of the funds required. Something of this sort has already started. In 1981, more than three hundred scientists, engineers, rocket technicians, and ex-astronauts banded together in the "Mars-Mafia." They have been agitating ever since for a manned landing and are working on ways of privately funding part of a plan to land by the year 2000.

The ship, with its circa 1995 state-of-the-art design, will make the shuttle and its accommodations seem old-fashioned but not unfamiliar. It will be designed for a larger crew than the shuttle can hold, and the crew will be aboard nearly twenty times as long as on the nominal thirty-day space shuttle missions.

The interior will therefore be designed with long-duration flights in mind. Since the amount of time spent in actually piloting a ship to another planet is small (most of the navigation will be done by computers), the ship's flight station may not be large. There will be seats for the command pilot, surrounded by the usual details of a flight deck: instruments, digital read-outs, switches, levers, and the inevitable computer screens.

The life-support system for a Mars mission will be the same as that used for the space station. The air and water would be recycled, but food would not be. Near the flight deck will be a mission station, which will be used during the orbital operations over Mars. It will contain the scientific packages and electronics that will narrow down the choice of a landing site.

The crew's quarters may be not unlike an old whaling ship or a modern yacht: full of nooks and crannies, cabinets, and nets in which gear and equipment are stowed. And as on ships everywhere, the space will be limited and each member of the crew will pack only one large bag for the voyage. There will be an extensive and potentially embarrassing security investigation of baggage. No one will be allowed to take stamps to Mars for cancellation or anything else of equivalent commercial value — unless, of course, it is authorized by NASA or some other government agency.

The galley will be quite roomy, with a pantry, an oven, a dishwasher, hot and cold running water, and a dining table. The oven, operating in a near-standard Earth atmosphere, will be able to heat a meal to nearly 200° and hold it at that temperature or slightly below for warming or reheating. The crew can sit down to breakfasts of orange drink, peaches, scrambled eggs, sausage, and cocoa — most of it reconstituted, but not all that bad. Dinner might be shrimp cocktail with sauce, beefsteak, broccoli au gratin, and drinks. They can even have a Coke, since the Coca-Cola Company has, in cooperation with NASA, designed a special can that can be used in space.

Electrical power for the ship could come from solar arrays, using sunlight to power the craft. The ship could also use fuel cells that have hydrogen and oxygen stored at extremely low temperatures. The fuel cells would not only produce sufficient electricity to power the equipment (the same technology is currently used on the space shuttle), but the by-product is, somewhat wonderfully, water. Thus all of the water needed for the voyage could come from the chemical conversion of hydrogen and oxygen to electricity. Extra fuel must of course be carried to provide the water; the extra weight is expensive but unavoidable.

While there is no particular purpose from an engineering standpoint for having these, there may be three small forward-facing viewports in the flight deck — a window. Even when the destination is millions of miles away, we like to see where we are going. And once in Mars orbit, the crew will be able to look out and see the great globe of the Red Planet hanging against the anthracite backdrop. The window will be made of "space glass," developed as a by-product of experiments in glass manufacturing aboard the spacelab and the space station. Set into the metal skin of the ship, it won't be very different from the windshield of a modern airliner or the windows of the space shuttle.

Computers aboard will have to be reliable, but since electronic components in space have been working for more than ten years (the *Voyager* spacecraft is expected to accept its command to photograph Neptune in 1989, more than ten years after it was launched), there should be few problems in designing a computer for a Mars expedition. Should a primary computer fail, there will be a worried moment, but the back-up computers should kick in immediately. Up to the point of no return, at which it becomes impractical to return directly to Earth, the ship could probably make it back home safely even if all the computers were to fail.

Once past the point of no return, the only option is to perform the "once around abort" maneuver and return to the home planet in an orbit that would take as long as the original trip. Unless there were some terrible emergency, the crew might as well finish the voyage to Mars.

The first mission would need a lander, some modern version of Wernher von Braun's 1947 "landing boat" or an oversized version of the *Viking* lander. A lander for Mars would not be unlike a lunar landing vehicle; even one of the antiquated LEMs from the Apollo program could make it to the Red Planet by sunrise 2000 if an appropriate ship to transport it to Mars orbit were available. Designing one from scratch, however, would be much better in the long run, and doing so using developments that will come about as we build the space station will make it relatively inexpensive.

A lander can be designed that would house a crew of three for a thirty-day stay on Mars, based on the use of scaled-down *Spacelab* technology. A mothership can be put together with currently available hardware or hardware created while building the space station. An Earth return vehicle can be similarly constructed. The total package for the first exploration of Mars would cost about $38.5 billion. The development costs of the space shuttle were about half that; the old lunar Apollo program would cost more than twice that in today's dollars.

The mission requires using the atmosphere of Mars as a "brake," which would save fuel costs, but "aerocapture technology," as the atmospheric braking is called, is nothing new in theory and it was used to some extent in the *Viking* landing.

One of the reasons scientists and engineers have proposed a spaceship that could be spun to provide artificial gravity is that we are unsure of the human factor. Are we up to a trip to Mars physically? The voyage would be six to nine months. Could we stand month after month in a zero-g spaceship and still be functional when we got to Mars?

People in space have had something of a hard time of it. For one thing, in the first few hours in zero-g, fluids redistribute themselves toward the head. This causes a lack of thirst; the brain assumes it has enough fluids even when the body does not. Blood plasma drops; the number of red blood cells — the cells that carry oxygen throughout the body — also drops. The heart, upon which everything depends in space as much as anywhere else, develops a major case of lassitude and shrinks slightly. It doesn't have to work against gravity any longer and it takes a break, slowing the pumping rate.

The body's muscles, especially in the legs, lose tone, elasticity, and strength. For every month in zero-g, the bones of the body lose minerals, particularly calcium, at the rate of about 0.5 percent a month. This is a serious calcium loss, but it may occur only in the first few months of the flight. Valeriy Ryumin, a Russian cosmonaut, has made two concurrent flights of six months each — a year in space — with no apparent ill effects. His "flight" was in zero-g.

On the other hand, if calcium loss is continuous throughout the flight, we might arrive on Mars with very brittle bones. This might not be a big problem in the lower Martian gravity, but it could be serious on the return to the normal gravity of

Earth. A solution does, however, exist: there are drugs that slow calcium loss, now used on bedridden patients and those in coma. Extended bed rest is not unlike the weightlessness of spaceflight, at least as far as the bones of the body are concerned.

The experience of living aboard a space station from 1990 onward will provide very useful information. We will find out not only about living in space for long periods and what effect that has on the human body, but also about what effects radiation might have over a long period of time. In the space station, crew members will be exposed to about the same radiation levels we expect on a voyage to the Red Planet. As we have discovered since 1945, it is the total dosage that counts, not the actual amount of exposure at one time unless that amount is in itself lethal. The current federal guidelines indicate a maximum lifetime exposure of 300 rems. A round trip to Mars would produce about one-fourth the federal maximum if there were no solar flares; a solar flare could raise the total by another 40 rems. Presumably OSHA would regard the voyage as an acceptable working environment if the astronauts involved made only one trip in their lifetimes.

Despite the apparent Russian success in long-duration flight studies aboard their space stations, whether the first expedition will be a zero-g flight or in a somewhat more complicated spacecraft that is spun for artificial gravity is open to question. Artificial gravity would make the trip easier on the crew; some of the same physical effects of zero-g would still be there, but vastly reduced. But a spaceship with artificial gravity is a somewhat more complex vehicle and one with which we have no experience.

In any event, life aboard the space station will provide many of the answers for a Mars voyage. Most scientists think the long flight would be not unlike a cruise on a nuclear submarine (that doesn't surface for six months) or a long stay aboard a *Sealab* project on the bottom of the ocean.

What can be said now is that boredom may be the largest psychological occupational hazard. It will be a study in a rare form of isolation — millions upon millions of miles from home, and yet in constant touch by radio, computer, and other electronic means. Although some people involved in planning Mars missions think there will be plenty of things to do — engineering checks, scientific observations, psychological evaluations, and a hundred others — there is a suspicion that their opinions may stem more from their love of the engineering point of view than from a realistic appraisal of the situation. After all, not everyone is as excited by scientific data as scientists are. People, even great space heroes, get profoundly bored. But there are ways around boredom, and long-term experience at the space station will provide clues on how to keep an expedition psychologically healthy.

In addition to providing experience in a dozen major fields — all of which are applicable to our planning — the space station will produce a large pool of experienced astronauts, scientists, engineers, and others who will have long-term space experience. It is from the crews of the station that the people who go to Mars will be selected. They are alive today, and their children may see them walking around on the Red Planet on the six o'clock news.

With all of this potential over the next few years, we are no longer at a time when talking about going to Mars is fantasy. It is not even politically unacceptable. It is almost within our grasp: shuttles, stations, rocket-spacesuits that can be used at construction sites, space tugs, and orbital transfer vehicles. The foundations of the voyage will be laid by 1992. All we have to do is take advantage of those foundations.

But to do so, we must first have a thorough unmanned

exploration. Viking was informative, but there's always more to know, more detective work to be done before we leap. There is no such thing as too much information about a planet on which we intend to land human beings.

We want to know more about water resources. We'd like to know more about the quantity of dust in the atmosphere: the extent, duration, and origin of the great storms that sweep the planet. We need cloud measurements and some further knowledge about the frosts that come to the surface.

All of this can be done with relatively cheap and simple spacecraft launched from Earth orbit out of the cargo bay of the space shuttle. Cost estimates of a water mapper run about a tenth as much as the old Viking mission. A surface mapping effort would also be extremely useful (we now describe the height of Olympus Mons as "somewhere between 18 and 27 miles," which is hardly precise).

NASA has funded a fairly complex geochemical orbiter for Mars to be launched sometime in 1990. The European Space Agency has designed the Kepler Mission, an orbiter intended to fill the gaps left over from the Viking exploration. Either or both of these should provide us with enough information for planning a successful manned effort. There will undoubtedly be some unknowns; that's why some astronauts are still classified as pilots.

At a million miles out, five days into the mission, Earth will look the size of the moon as seen from the space station; it will grow smaller with each passing minute. Half of it will be in sunlight, a waxing half-Earth. The moon, only one-quarter as large, will be a disklet in black.

The crew will settle down to the long voyage. Some of them may have a full mission's worth of practice time aboard the space station, but most of them will not have (after all, that would mean more exposure to radiation). If the trip is in zero-g, they will begin to look strange to each other. The redistribution of fluids toward their upper bodies will have some interesting effects. They will feel a sensation of fullness in their heads and a slight nasal congestion. They will see the puffy faces of their companions (and their own) in the polished metal mirrors of the ship. Glass mirrors are entirely too dangerous: a shattered mirror in zero-g would be nearly as lethal as a hand grenade.

Their legs will grow thinner, an exaggerated condition that will make them look as if they have crow's legs. On the average, their chests, biceps, and necks will grow an inch in circumference; their thighs and calves will lose an inch.

They will feel a slight loss of balance because additional fluid will be in the inner ear; blood will pool in their chests, reducing lung capacity. Body fluid will settle in their backs, where it will be absorbed by the spongy disks that separate the vertebrae. The disks will grow thicker and the vertebrae will be pushed much farther apart than normal. This will make the crew feel taller — and they will be too, as much as two inches in some cases.

With their pulses averaging about twenty beats lower than on Earth, the crew will move about their compartments without using their feet. For billions of years we have been used to gravity. Everything about our bodies reacts as if gravity were present. In its absence, things grow strange for us. Even if some enterprising astronaut tried to walk on the deck, the result would only be forward motion until he or she hit something.

Their flight suits will have plenty of pockets. In zero-g every little item must be in a pocket and snapped in place; otherwise the object will float out of the pocket and have to be recovered later. The flight suits will also be loose-fitting. Partly this is for comfort. Partly it is for ease in dressing. Putting on shoes in space is a grand trick; you use your stomach muscles to pull a foot up close, so you can work on a sock or tie the laces.

The crew's shoes may have aluminum soles covered with rubber onto which they can fasten metal cleats. Some parts of the ship will have metal triangular grids in the "floor," so the crew can "stand" from time to time and not spend the entire voyage drifting about in a dreamlike, disembodied way, moving as if they were underwater. Although there is no up or down to them, for purposes of sanity the floor and the ceiling will be arranged with the same general perspective as on Earth. Astronauts tend to stand on the floor when possible, since that is our most familiar sensation. This gives them an artificial sense of up and down, or what astronauts call the "local vertical."

For much the same reasons of familiarity, there will be cabinets aboard. The cabinets will have small shelves lined with Velcro on which objects can be stored. The Velcro strips will keep the objects in place so they will not come shooting out the minute the cabinet is opened.

Eighty days out, with the crew's first tendrils of boredom thickening, Earth will be more than ten million miles away and will pass between the ship and the sun. If the circumstances are exactly right, the crew will be able to see an astronomical oddity, a genuine "transit" of Earth and its moon across the sun's small disk. They would see a tiny black dot amid the eternal flames of the solar corona; the dot would move from one side to the other, to be followed an hour later by the smaller black dot that is the moon. The crossing would last for eight hours.

Despite constant communication with the space station,

or perhaps even because of it, the crew may feel isolation descend on elephant feet. Personality problems may not be ironed out quite as quickly as psychologists on Earth and aboard the space station think they will. Communications with the station will eventually become annoying, since as the distance increases, the time lag also increases. Radio communication travels at the same speed as light. That's about the fastest thing around, normally instantaneous. But over a great enough distance, even light takes a while to get there.

The time lag in communication is only slightly troublesome at the distance of the moon from Earth — the conversation just seems slower as each side pauses for the slight delay. At the distance of the expedition from the space station, however, even after a few weeks, the lag will be very noticeable. It will grow to nearly twenty minutes when the ship is near its Mars orbit. Keeping up a conversation will eventually be a bit like a chess game by mail — very tedious unless there is something else to do. After the halfway point in the trip, it would be extremely difficult for crew members to have an argument with the space station, since in the time needed for the angry words to fly through space, the parties would have too much time to cool off.

They will spend their time doing engineering checks, and they'll check on the lander avionics and supplies in the cargo hold. Most of all, they'll try to avoid boredom. Card games will be played with thin metal cards and a magnetic table, or perhaps with thin Velcro-backed plastic cards and a Velcro-covered table, depending on the game being played. Shuffling, in either event, will be extremely difficult, leading more often than not to the old children's joke of "fifty-two pickup." Some crew members will play chess or checkers, either of which is easier in zero-g than trying to deal with decks of cards. The pieces will be metal and the boards magnetic. They can even play games with the crew of the space station until the time lag makes it relatively impractical for a game lasting less than a few days.

The most entertaining event on the long trip may be watching movies. A large collection of movie tapes will be stored in the crew's quarters for replay over television monitors. The rush of videotape technology should make possible microcassettes of movies played on a machine not much larger than a portable tape recorder. It will not be much different from having the neighbors over for a night of horror movies except that the audience might watch the movie from unusual positions; there would be no sense in having "seats." Everybody would just latch on to something to avoid drifting about in the slight currents from the air recirculation ventilators. Popcorn would not be out of the question: self-contained popping bags could be placed in a microwave, although the steam and oils from popping corn would not be easy to deal with in a zero-g situation.

Earth television could be piped aboard too; the time lag would not be a problem since the programs would be taped aboard for later showing. The six o'clock news would be less than half an hour late, even when the ship neared Mars.

Exercise will be mandatory if the crew is to survive even the weak gravity on the Martian surface. After only a few months in space without constant exercise, the crew would have trouble adjusting to the Martian gravity and it would be difficult for them to set up the expedition base. The exercise hardware will not be markedly different from that found in any zero-g environment, such as the *Spacelab* of 1984 to 1995 or the space stations. There will be bicycles, rope and handle exercisers, and treadmill devices in a compact form.

For general housekeeping the crew will use small, handheld vacuum cleaners; these will be used to remove unwanted particles from the air and to clean the galley. Trash will be stored in the galley and eventually shot into space through a hatch, leaving a trail, if anyone wanted to follow it, all the way to Mars. The trash would retain the velocity of the space-ship for some time and then fall slowly behind. If the ship were aimed perfectly, the trash might wind up near Mars several years later, where it would burn up in the atmosphere. In the most likely instance, however, the trash would miss Mars and go off into space, to be wondered about sometime by whatever alien civilization picked it up.

The ship will be equipped with bunks not unlike those aboard a nuclear submarine. The bunks will have draw curtains like those on the old Pullman train cars or aboard first-class transatlantic airplanes. The bunks will be fitted with restraining straps; otherwise the completely relaxed sleeper would drift out of the bunk and into the crew's quarters on whatever air currents happened to be moving late at night. For those who do not want to sleep in a bunk, there will be space bags, which are now used aboard the shuttle. They resemble, more than anything else, morgue body bags but are made out of woven material instead of plastic. In zero-g, where there is no sensation of weight, you get into your bag, zip it nearly shut, and go to sleep with the bag hanging from a hook; a restraining strap holds the bottom to a bulkhead.

The crew will bathe in plastic bags. Water behaves in strange ways in zero-g — it's quite undisciplined and unin-clined to remain in a container — so bathing in space is not much like bathing on Earth. The lack of gravity causes the water to go everywhere, sometimes in a big globular form looking like a water bubble in oil, and sometimes in a spray of tiny droplets that stay in suspension in the air.

The *pissoire,* that ubiquitous cabinet with which any house in space — moving or not — must be equipped, is a relatively simple affair. It will not differ from those aboard the space station; in fact, space toilets are not very different from those aboard airliners, except for the special provisions required by the absence of gravity. Wastes are separated from the body by a vacuum flow, and the contoured seat has a seat belt so the user can thwart Newton's Third Law and remain in the correct position. For males, there will be footholds and handholds as well.

Arthur C. Clarke is supposed to have once remarked that space travel would bring forth new forms of erotica — and about time, too, he added. *Whether* is not really the question; *how* is more to the point. Leaving dusty notions of Victorian morality aside, both male and female astronauts will go to Mars, and during the voyage they will certainly find time to make love. Pertinent research will probably have been done aboard the space station long before the Mars ship leaves, since the station is expected to have mixed crews as well. As far as we know, sex in zero-g has not yet been tried; it is relatively impractical to experiment on Earth (true zero-g can be achieved in aircraft flying a parabolic course, but the effect is very short). About anything that might have occurred on a space flight, no one is talking.

Bodies in zero-g do not easily remain in contact. Any movement, any physical action, is followed immediately by an opposite and equally strong reaction. This is called Newton's Third Law, the principle on which rockets fly. It means that any physical movement between two people would generally result in both of them flying around the room. One solution is to have a separate cubicle aboard ship with a variety of hand- and footholds, the use of which would require considerable athletic ability. Another possibility would be to design a form of oversized girdle for two, an elastic (but not too elastic) web.

Love and the zero-g will require pretty fancy house-keeping around the area in which it is done. In addition to being used to gravity for all of human history, we are used to most things remaining in their place — like liquids. In space, a globule of sweat does not drip off the nose to the deck. It flies off on a little orbit of its own and remains in suspension until someone does something about it with a hand-held vacuum cleaner or similar device, or until it is swept into a ventilation duct.

If the voyage to Mars is an artificial gravity flight, it will be more pleasant, but not without its own hazards. Several on-paper designs for Mars missions have been conceived using modules that can be spun. At three revolutions per minute, the living quarters of the interplanetary spacecraft would have about one-quarter of Earth's gravity. With a 0.25 artificial gravity, the crew would feel some normal sensations of weight and balance; but the gravity is sufficiently less than Earth's, so readjustment would be required. Things don't go flying around in lesser gravity, but they do not stay in place as well as they do in an Earth-normal gravity environment. Movements would have to be controlled so that the experiences (and powerful muscles) of Earth did not leave the crew forever feeling like bulls in a china cabinet.

But with one diversion or another, zero or artificial gravity, the months of the trip will pass, and if the crew is lucky with midcourse corrections, the arrival will come within a few minutes of the time predicted long ago by the computers aboard the space station. Such accuracy is old news in the space business: one *Pioneer* spacecraft was only a minute off after traveling more than 500 million miles. This was nearly three decades before our spaceship to Mars will be crossing the black void.

Once the ship is in orbit around the planet, two members

of the crew will crawl through an entry hatch/airlock into the cargo section. They will activate the lander systems so the little craft can be prepared for launching. When everything has been checked out, the lander will be ready to power down to the red world below.

There may be an extremely unhappy female crew member originally scheduled to be one of the landing party — a pregnant crew member. A Hohmann orbit to Mars is about as long as a pregnancy, and other transfer orbits are at least six months. There would be no possibility of returning to Earth for the birth. The infant would be the first real spaceman or spacewoman worthy of the name — in truth, a child of the universe.

A pregnancy in space poses some problems that will probably not have been resolved by experience aboard the space station. After all, if a pregnancy did happen at the station, the parents-to-be would be near Earth and could choose the home planet for most of the pregnancy and the delivery. The most serious problem of a pregnancy in space is radiation. Cosmic rays break chemical bonds when they go through tissue, which can damage cells and lead to cancer, genetic mutations, and mangled chromosomes. Moreover, iron nuclei, which are included in cosmic rays, cause damage to nerve cells of the brain and spinal column. These cells in humans do not reproduce, so the loss is permanent. On the face of it, a space pregnancy sounds like a bad idea.

The expected radiation dosage on a trip to Mars, 40 rems, is eight times the annual amount allowed by the Atomic Energy Commission for adults working in the nuclear industry and is much higher than the amount recommended for children and pregnant women in the general population. But the figure of 40 rems is just an estimate; it may be less on an actual trip, especially if *Mars-1*'s fuel tanks are arranged so that the crew's quarters are shielded by the tanks from radiation. And some radiation studies have shown that even a dosage of 50 rems a year has no detectable effects. Women who have lived all their lives in very high altitudes — for instance, Lhasa, Tibet; Bogotá, Colombia; or Nepal — have accumulated total dosages from natural sources higher than the dosage that would be received on a trip to Mars, and they still get pregnant and have normal births.

In any event, a pregnancy aboard *Mars-1* would be a case of *vogue la galère:* press on regardless. The problem for the female crew member is that in her condition she no longer makes a good candidate for a planetary landing. For the birth itself the doctor would prefer the one-third gravity (a third of Earth's gravity, that is) of Mars over the zero-g of space; but because of the relatively high g-forces of a landing, the doctor will almost certainly order the unhappy crew member to stay aboard ship. The doctor will stay aboard ship too, of course, as will a crew member who will be the ship's liaison with the party that is about to leave for the planet's surface.

4 WE ARE THE MARTIANS NOW

THE BLACK VOID outside the viewports of *Mars-1* will be dominated by the great swollen globe of the planet as the ship orbits Mars. The lander, dormant for nearly nine months except for engineering checks of its systems, will be alive with working electronics. The surface crew will make their final checks and the order will be given to separate from the mothership.

Detached and free, the lander will remain close to the ship until the landing sequence has been checked one last time; the word will be radioed to Earth that the lander is departing Mars orbit. On January 26, 2002, the landing craft will have ignition on its engines. Slowly at first, then much more rapidly, the flight path of the lander will diverge from the ship's and bend down toward the planet. The crew will feel the first strings of random molecules that pass for the upper atmosphere. At least they will *think* so: it will register on their instruments, if not their human senses. Spread out before them from a very high altitude will be half a hemisphere of the Red Planet.

To their left will be the giant Tharsis Ridge with a string of volcanoes: Arsia Mons, Pavonis Mons, and Ascraeus Mons. Just visible on the northwest flank of the Tharsis uplift will be Olympus, the largest volcano in the solar system. All of them dwarf Mauna Loa, the best that Earth and the Hawaiian Islands can offer. The great scarp surrounding Olympus Mons will be sharp and clear unless there is one of the rare but thorough Martian dust storms.

In the upper part of their view will be the intricately patterned fractures, a mud-crack network of interconnecting valleys, called Labyrinthus Noctis. It is the western end of the great rift system that threads eastward, becoming deeper and wider until it is the Valles Marineris, a system that cuts a series of grooves a quarter of the way around the globe. As it goes east, the system of chasms gives way to broad, riverlike valleys that meander northward to Chryse. The Great Rift Valley of East Africa is only a comparison for this cracked, rugged vista.

On their right, east of Lunae Planum, somewhere over the horizon, will be the place where the *Viking* spacecraft will still be standing amid the rocks so familiar to scientists on Earth. A blue-shaded sky, a result of early morning ice-crystal fog, will still shine on the pitted and dust-covered little explorer that fell silent so long before.

The lander will drop lower; the curve of the Martian horizon will disappear; the landscape seen from the viewports will narrow. The view will contain the vast bowl of Syria, Sinai, and Solis Planum, the cratered plains that flow from about 60°W to 100°W. South and slightly east of Solis Planum is the old telescopic feature called Solis Lacus, the Lake of the Sun, also known as the "eye of Mars." Dark basaltic bedrock overlain by a veneer of thin sediment covers thousands of square miles.

In 1867 Richard Proctor, unable to see the true nature of the Martian surface through his telescopes, named this part of Mars after the Italian astronomer Father Angelo Secchi. He called it the Secchi Continent, south of the "Huygens Marshes." There is very little similarity between what Proctor and Secchi saw in their telescopes and what the surface of Mars looks like from high altitude.

Lowell had seen this thinly cratered, basaltic terrain as a center of canals. Ambrosia canal went south, connecting with Acis. Bathys went southeast, connecting with Astrae Lacus. There was a network of them going north from Solis Lacus: Corax, Coprates (a double canal), Tithonius, Eosphorus, and Aesis. The great double ran northward to the oasis of Tithonius Lacus. Lowell's 1895 map had 185 canals; more than a dozen were located in the eye of Mars. In the end, when the spacecraft cameras had done with his memory and his dreams, all he received for this enormous effort of hours upon hours at the telescope and hundreds more poring over mythological figures and references was an honorable mention. Lowell crater, a relatively recent, central-ringed basin about 120 miles across, lies at 81°W, 51°S, quite a bit south of Solis Lacus and a little east of what he thought was the canal Acis.

The spaceship's lander will drop down with a little bucking from high atmosphere turbulence, and an imagined rising moan as it hits what passes for heavy atmosphere and sinks to less than 20,000 feet. The first manned landing on Mars will be less than two minutes away. The terminal descent engines will fire; the Martian sands will spread outward in a swirling mushroom under the rocket nozzles as the lander comes down. The sands of Mars aren't really that: they are fine-grained silt and clay. But to us, they will always be sands, no matter what the geologists call the thin veneer of bright, fine-grained dust that covers what we will forever call dunes.

The weather at the landing site will be mild — what we would think of as a sunny but extremely cold day. The early morning will have been nippy at $-122°F$, warming to $-22°$ for the day's high. Dust particles floating in the thin atmosphere will scatter the sunlight to paint the sky pink. There will be light winds shifting from the east to the southwest at a maximum speed of 15 mph. Phobos will be a dim half-moon high in the sky, perhaps invisible behind a thin cloud. The pilot of the lander will spend fourteen seconds drifting over the landscape at 50 feet to avoid any string of boulders or rough terrain that might damage the craft.

There will be two preselected landing sites: Syrtis Major and Solis Lacus. Syrtis Major is the oldest known feature on Mars; Christian Huygens made a crude sketch of it as seen through a telescope in 1659. It is a wind-streaked plain surrounded by blasted craterlands on the north, west, and south. To the east is a huge basin called Isidis, where a million or two million or an untold number of millions of years ago, a piece of the asteroid belt plowed into the Martian surface and hid there. The great plain was formed when the asteroid fractured the planet's crust and magma flowed over and collapsed the surrounding terrain.

The second landing site, the one that will be chosen on this January morning in 2002, is the northeastern corner of Solis Lacus. Lowell called it Nessonis Lacus. In 1929 the French astronomer Antoniadi named it Vestae Depresso, for the goddess of the domestic hearth. Hers was the temple where the holy fire burned forever. On Earth, the equivalent location is a marsh in Thessaly. The landing site lies at 85°W, 22°S on Mars.

The expedition will hit dirt a few seconds after ten A.M. local time. The success of the landing will be radioed to the orbiting mothership. Meanwhile, "mothership" will have taken on a new meaning for the Mars astronauts: the pregnant crew member is feeling the first labor pains.

Once the lander is on Mars, the crew will rest until the next day. They will need some time to get used to gravity again, even the weak gravity of Mars, after almost ten months of zero-g. Their coffee cups will no longer stay in midair; they'll fall and break. Pencils left hovering above a console will drop onto the instruments. The crew will have to practice walking again in the severely limited space of the lander. Even if the trip has been in artificial gravity, they may wait a day before emerging.

The exact time at which a ribbed foot will enter the fog snakes of an early morning ice-crystal layer and stamp a tread into the rusty, iron-rich soils of Mars — in this case into a dark patch of ground that would have been called a mud puddle if it had been anywhere else in the solar system — is a matter of semantics. The "day" on that patch of ground is forty minutes longer than on Earth. But the clocks aboard the ship will have been programmed to count each Earth minute plus a fraction of a second as a Martian minute, so the time of the landing will be perfectly good local time.

It will be early morning of the second day when the crew members begin preparations to step onto the surface. The light will sharply delineate the scene. The accumulations of sediment among the jutting rocks of the landscape will be more reminiscent of Earth's sand dunes than anything else. Most of the rocks around the lander will be small, less than a yard in diameter; irregularly shaped smaller pieces will be scattered here and there. Some of the rocks will be heavily eroded by the wind. The smaller stones and pebbles around the landing site will be somewhat like Earth's caliche in the American southwest. The caliche-like surface of Mars is caused by upward migration of water through the soil that precipitates salts after it evaporates.

The horizon, which always appears close on Mars, will be less than 2 miles away. On the skyline will be dips and humps that represent parts of raised craters. The picture from the lander will be slightly out of perspective: the landing was not entirely perfect. The lander may be tilted a few degrees away from the vertical.

The surface crew will put on their environment suits and open a communication channel to Earth via the orbiting mothership. A sequence of switches will be thrown and outside the lander a honeycomb aluminum ladder will slide out of the ship and angle down to the surface of the planet. Another sequence of switches and the outer hatch will open into the Martian sunlight.

Poised on top of the ladder, breathing easily and freely from the life-support system of the environment suit, the first human being on Mars will not notice the cold. He or she will be thinking of the moment, a few seconds later, when the first foot will step out onto the Martian sands. What might he or she say? When it comes down to the real landing, the first words may be something quite different from what we suppose today. But it might be appropriate to make the remark that Ray Bradbury made so long ago when the little *Viking* spacecraft first touched the soil of the Red Planet: "We are the Martians now."

Slowly the first human will descend the ladder. One foot steps into the dirt, leaving a footprint amid a little chuff of dust that rises into the cold air. The other foot comes down and makes a second footprint and another small dust cloud. We have arrived.

The new Martians will feel strange in their suits. Their breath will hiss in and out, rasping slightly; the sound will mix with low noises from the life-support system. The helmet will provide a clear 240-degree view. The explorers won't be able to see what's behind them without turning; but since they won't be expecting anything to come up behind them, it won't matter. The helmet will be tinted against the harsh light of Mars, but the tinting will be relatively slight, not the highly reflective silver shielding used on the moon.

Their helmets will be full of conversations between the surface party and the mothership. No communication with Earth in a direct sense is possible: the home planet is more than twenty minutes away by radio. The Martian desert will lie before them, timelessly etched on their memories. The sands will have a pattern of footprints for the first time.

As noon approaches, the light will go flat, the shadows of the rocks will disappear, and the landscape will look much like the Arizona desert. Even at midday, the temperatures will be low, below freezing. The sky will be the pastel-pink of Mars, turning in places to a deep blue-black. There may be thin clouds or perhaps a dense section of clouds filled with dust. Thick cloud banks are rare on Mars, but the sky is often partially filled with thin cirrus formations at high altitudes.

They may see a most unusual phenomenon: a small puff of gas from one of the still-active volcanic vents. While there is disagreement, some serious watchers of Mars think minor volcanic activity is still present. *Viking* photos taken in 1978 showed tiny clouds whose rapid changes in a few seconds suggested active vulcanism.

Looking into a sky that holds a sun much smaller than the one seen from earth, the first men and women on Mars may see one or even both of the tiny moons traveling their strange courses, making swift changes of phase from full to half to new. Those moons will be very important to the future of mankind on the planet.

Above the close horizon, at the limits of their vision, the crew may see a Martian dust devil. A dust devil on Earth's deserts can vary from a wind-infant to a gigantic, swirling storm that can pick up objects weighing several pounds. Somewhat predictably, the Martian counterpart is a monster that, like the Martian volcanoes, dwarfs anything on Earth. On Mars, wind is the major force that changes the face of the planet. The tornado-like columns of dust can rise to a height of $3^1/_2$ miles above the dunes and parched plains. When the sun is high in the sky near the equator, the soil heats rapidly. If the temperature is right and there are enough loose surface grains, a great dust devil is born. Inside the swirling maelstrom, the dust grains are swept around at greater and greater speeds. The dust devil towers above the terrain and can move with the generalized winds at high speeds.

The second day on Mars will draw to a close and the crew will return to the lander, dusty, tired, and dirty. The sun will drop lower in the sky; and in the thickening Martian dusk, the cold will return to the sands. Night will fall around the lander and the sky will turn black, sprinkled with stars. One of those stars, a relatively bright blue one, is the planet Earth.

These first explorers will be roughing it. As in the first landing on the moon, they will learn as they progress, fail as they go forward with their work. Part of the excitement of the first expedition will be the pure human adventure, but there will be science too, and engineering. They will look for water and test it. They will survey the landscape and compare the rocks with those photographed so long ago by *Viking*. They will set up their camp under the strange sky and survive with their closed-loop life-support systems. Most of the time they will walk around clumsily, looking as if they had escaped from the set of a bad science fiction movie.

On their third day on the planet, the crew will begin unloading cargo. The first crate will be a simple solar cooker. Powered by the sunlight hitting the red world — less than on Earth, but still a considerable amount — the solar cooker will heat several hundred pounds of soil at a time. It will work only when the sky is not cloudy, but it is not cloudy most of the time on Mars. The Martian soil, when heated, gives up oxygen that the explorers can use during their stay on the planet.

An additional solar cooker will reduce large chunks of permafrost to reasonably useful water. The taste may be a bit fierce: from the minerals expected in the soil, it can be predicted that the water will be extremely hard and perhaps salty. The water can also be a source of oxygen, although not as good a source as the Martian soil. If a trace of sulfuric acid is added to the water (the acid would need to be brought from Earth, although it would be possible to manufacture it from native resources given the right equipment and enough time) and it is then subjected to electrolysis, oxygen would be released. After two or three days, the area around the landing site may look more like the aftermath of a raid on a moonshiners' paradise than the home of a scientific party.

When they have some supplies of oxygen and water coming from native resources, the crew will use small charges of explosives to scoop a fairly deep hole in a nearby hill and set up the foundations for the first house. A house on Mars is better buried underground, or at least partly underground. This provides some protection from the strong cosmic ray bath, as well as the strong ultraviolet flux caused by the thin atmosphere. If the fine-grained, powdery sands are used to fill in around the house, the insulation is very effective. The structure of the house will be an oval of thin Mylar — a dome set into the hole carved in the hill. Mylar is one of the cheapest building materials available and one of the lightest to transport.

The water electrolysis plant will provide more than oxygen: the other component of water is hydrogen, which can be caused to react with the carbon dioxide of the Martian air. The resulting smelly mess, methane, can be used for heating the house, although the house can also be heated by ordinary solar means.

The environment suits will be uncomfortable. Engineers who design spacesuits should live in one for a few days, in the same way architects or interior decorators should be made to live with their creations. The crew will hate them. The first Martians will be irritable, tired, and scratching when they are finally able to live in the first dome. Manual labor in the environment suits, even with a break inside the lander at night, will be an easily forgettable experience.

After a few days the crew will begin work on a second dome; this will be an experimental greenhouse. As we know from experiments that have been performed on Earth, plants can be grown in *approximations* of Martian soil. The key word is "approximation." The real Martian soil amounts to a sterile, silt-clay substrate with some exceedingly uninteresting chemicals in it. While there is a trace of nitrogen in the atmosphere, gaseous nitrogen is not useful to plants. They use nitrogen that has been "fixed" in the soil by bacteria and then released as nitrate. There are no N-fixing bacteria on Mars, as far as we now know.

The greenhouse will have trays filled with soil that has been modified to remove some of the potential disasters it might have for plants. Using local water that has also been slightly treated, the astronauts will see what plants will grow.

The Martian air has twenty times as much carbon dioxide as our own, so plants are expected to flourish if the soil is properly prepared. The first experiments in the greenhouse may be with natural Martian air supplies (but at increased pressure), and the job of tending the greenhouse may not be popular: it will require suiting up, whereas work in the main dome, which will have an Earth-like atmosphere, will not.

Most scientists believe that plants will grow much taller in the lesser gravity of the Red Planet; how much taller the plants are and how radically different the growth pattern is will be two of the thousands of bits of data the first expedition will bring back to Earth. It is impossible to establish plant growth in partial gravity conditions on Earth, and information from the zero-g of a space station or the spacelab may not be a good indicator of how plants will grow in less-than-Earth-gravity situations other than zero-g.

Some recent studies of plants grown in approximations of Martian soil have led to the conclusion that a greenhouse with a "farm" surface area equal to about three football fields would yield enough food for fifteen astronauts. The first expedition's greenhouse will not, of course, be anywhere near that large; but it will not have to be because the astronauts will primarily use food supplies brought from Earth.

After a week the camp will take on its final shape. The greenhouse will have a workable atmosphere for plants and the trays will be seeded. The two houses, one for the astronauts and one for the scientific instruments, will be completed and ready to move into. The solar cookers and electrolysis plants will be working at full capacity producing oxygen and water. The lander will be abandoned as headquarters, to the crew's enormous relief. The lander as a living space could best be described by an articulate canned sardine.

Building the base will have been a physical strain. Everything on Mars will feel lighter than on Earth — one hundred pounds will feel like only thirty-eight — but not to muscles unused to work for nine months. Creatures with a musculo-skeletal system developed on a planet like ours would eventually find life on a light-gravity world a little too delightful. The muscles would atrophy. When the person finally returned to the home planet, the result would be extremely painful; the legs would not support the body's weight. Everything would feel as if it were in slow motion. Life would be like moving in syrup.

The members of the first expedition will hardly be exposed to the problem long enough, however; and in any event they will have to cope with the return trip. They will have devised exercises to compensate. But what of any permanent settlement on Mars? Fortunately, the solution is fairly simple: it's called exercise. More of it would be required than on Earth for the same fitness because twenty pushups on Mars are not equivalent to twenty pushups on Earth. Martian pushups don't have the same muscle-building effect.

The trick on the Red Planet is to exercise against something that is independent of gravity, such as a spring. Springs have the same tension on either planet; the same amount of muscular effort is required to move the spring the same distance in either place. Wrestling works the same way; the exercise is against a force, not the weight of the opponent. A wrestler could throw another one a fair distance on Mars, but pushing against another human's muscles is the same exercise and strain as on Earth.

The urge to explore beyond the immediate area of the landing site will be hard to resist. During the Viking program, scientists who saw Martian landscapes through the fixed camera-eyes of the spacecraft found that the desire to "walk" over the nearest hill was quite irresistible. There is no question that a manned expedition will want to do some exploring at a much greater distance than would be possible on foot. With the nearly Earth-like surface view of desert and hills there will always be the feeling that just over the next hill will be something spectacularly interesting.

Taking long walks will be difficult, and the amount of oxygen available in the body packs of the suits will keep crew members from going very far. They could carry enough extra oxygen for a long walk using carts — normally easy to pull in the low Martian gravity — but remaining in the suits for any great length of time will be ignoble at best. A walk of nearly twenty hours could be made. A standard "moon suit" from the old Apollo days would keep a man or woman alive for four or perhaps five days. The Mars environment suit would be at least that efficient. If someone chose to take that long a walk, he could cover about 10 miles a day at some risk. The suit's water supply is tasteless and so is the access nipple. The food, if it can be called that, is stored as a paste aboard the suit and would be universally acclaimed the worst in the galaxy. The most degrading aspect of the suit will be the diaper, unless some way has been found around that pretty little technical problem by the time the first expedition for Mars sets off from the space station. Fluids can be drained out through a one-way valve; the solids have to remain within the suit — and with the astronaut.

The people on Mars will be held to a minute-by-minute time line that will have been prepared aboard the space station and on Earth. It will be detailed, monstrously complex, and not terribly interesting after a few days. It will be the Martian explorers' lot, as it was that of the Apollo astronauts before them, to perform the job of scientists without being scientists. They will be asked for measurements that will lead to conclusions about the composition of rocks and soil when they would prefer to wander over yonder hill and see what's on the other side. They may not be very interested in whether the Martian igneous rocks contain xenoliths. But they will gather them all the same, and carefully label each together with a photograph of the survey site.

After four days of roaming around the landing site in suits, the crew will unpack their rovers. The rovers will be a sophisticated version of the lightweight vehicle used on the moon in the early 1970s. The Mars rovers will be derived primarily from a completely enclosed mobile geological laboratory that ranged over the volcanic craters of southern California more than twenty years before the landing on Mars. They will be self-contained, meaning that suits won't be needed. The passengers will be at least as safe in them in the alien environment as they would be in any undersea exploration vehicle on Earth.

The new Martians will move 50–60 miles a day. On short trips they can explore the old telescopic regions of Thaumasia and Bosporus. Thaumasia was named after Thumas, the god of clouds and celestial ghosts. If the explorers head in that direction, they will be driving into a relatively flat area that opens toward the craters Lampland, Lowell, and Slipher. Carl Otto Lampland was an assistant of Percival Lowell's at the Flagstaff Observatory around 1903; so was Vesto Melvin Slipher, who was famous, among other observations of Mars, for his work on the giant Martian dust storm in 1954. The storm, which drew worldwide attention, was shaped into a very distinct W that lasted in the atmosphere for weeks. So great still was the public fascination with the Martians of H. G. Wells that the less sophisticated interpreted the W as a signal that Mars was preparing for war.

If the crew members take the rovers and head in roughly the opposite direction, across the old telescopic features of

Eosphori Lacus and Phoenicis Lacus, west of the imaginary canal called Tithonius, they will run into the fantastic canyons of Mars, including Labyrinthus Noctis, which from orbit looks like a cat's cradle of deep grooves. They may be able to drive the rovers all the way to the lip of the canyons and watch early morning ice-crystal fogs break up and float away in the morning sun. The canyon bottoms are, of course, low areas on Mars; and there the atmospheric pressure is a little greater, perhaps leading to small regional hazes and fogs of ice crystals that form during the night, sublimate with the rising sun, and then condense again with the cool of an evening.

Many of the interlacing deep, canyonlike grooves of the Labyrinthus are six to seven miles wide, so the rovers will be stopped at the edge of the massive canyonlands. Skirting them would involve a great distance and might be almost as difficult as getting down the canyon walls. This region of Mars, while fascinating, is extremely rough. The area is blanketed by volcanic deposits, since the great volcanoes of Mars are not too distant.

Somewhere on one trip or another from the landing base, the explorers will cross the spot where the canal Corax should be. Lowell put it on his maps of 1895. It was named after a river that on Earth flows into the Black Sea. There will be nothing more than rocks and sand and sand and rocks; the permafrost layer may be almost on the surface in parts of the region. The crew will be traveling over what would be, if separated out and melted, a modest amount of water.

The Martian desert will be about as lonely as any desert can get. People can isolate themselves from almost everything for a time; they can adopt mental attitudes that cope with loneliness. But the body is not so easily sidetracked; it remembers the companionship of a populated world. The crew members will have been nine months in isolation aboard a spaceship. They will have exhausted themselves in building a base. Out there on that blasted heath on a marginal planet that is one of the greatest human dreams, they will feel the loneliness creep in.

Inside their little driving compartment aboard the rover, they will park under the bowl of desert sky. Breathing Earth-normal air, they will stare out the windows at the nearby landscape lit by the three floodlights on the roof of the vehicle. The jumble of rocks visible will be not very distinctive; the explorers may be rather tired of describing geological formations to their microcassette recorders. If they turn off the floods, they can look out the plastic dome windshield of the rover and see the faint light of the Martian moons, throwing dim illumination on the deserts. In the distance, unseen in the dark, is Olympus Mons and the volcano region. Above in that black bowl of the sky, possibly just visible to the crew in the rover if the orbit is just right and the distance from Mars is suitable, is the mothership that will eventually take them home. Up there in the grand silence, a baby will soon begin making its first loud wail.

As the new Martians watch that faint light move across their sky, they will feel singularly lonely. If that light were to disappear, their chances of survival would be nil. Despite the experimental greenhouse, the oxygen-manufacturing machines, the houses and domes, the water-generating plants, they are on the edge of survival. They could breathe the reconstituted Martian air for months; they would have heat and light; they might be relatively comfortable for a time — but the food would run out.

The eventual result would be more tragic, if that is possible, than the lonely death of Scott and his companions in a frigid Antarctic storm almost a century earlier. They died almost within sight of their next supply cache. On Mars, the expedition would have no cache if the mothership were destroyed. Cannibalism would be useless: it would only prolong the agony. The nearest supplies would be nine months away by rocket.

Still, in the silence of that desert, they will have each other for companionship. Humans need sounds, smells, voices, and touch as much as they need phosphorus and calcium. In the morning, they will move on and follow their exploration time lines. The night fears will be gone, as on Earth, with the coming of the sun.

The crew will range far enough from the base to bring back vastly different core samples. They will use Black & Decker drills not unlike those used on the moon three decades before. The drills can penetrate 20 feet, except in permafrost, where they bog down to one-tenth their normal speed. They will also be looking for a good crater, which in this part of Mars should not be hard to find.

Small craters in the area may show up from orbit by the streaks or wind-blown sand that sweep outward from the rim like a black fan. A crater is a natural core drill, a cosmic geological dig into the ancient crust. Material from inside a crater could tell scientists a great deal about the history of Mars and its geologic periods. Study of crater material, especially on-the-spot examination by astronauts and cameras, might answer the fundamental question of why there is no longer liquid water on Mars.

Much of this detailed exploration work is extremely important: the history of Mars is tied in with the future of Mars. If scientists know what processes were at work there in the past, they may be able to reverse or destabilize those forces now or during a later colonization effort. This might result in the planet becoming more like Earth, which would be of obvious benefit to us.

Eventually the crew's rovers will reach their limits. The next expedition will have a prototype airplane and a host of other wonders. The next expedition will build a permanent settlement. There may even be two expeditions and two towns; they could use the airplane to commute.

While the exploration is taking place on the planet, another interesting human event will be taking place in Mars orbit. The birth aboard the orbiting spaceship may be quite normal in every respect. Experiments, if they could be called that, carried out in the 1970s during the search for more natural childbirth than was natural, showed that a water environment such as a swimming pool causes few problems; it is a fairly good approximation of zero-g.

The main difficulty is that babies are no longer gill-equipped and find the emergence from the womb into water disturbing, evolution having left them expecting something like air to breathe. Experiments with zero gravity aboard the old *Skylab* in the early 1970s also showed that certain species adapted quickly to the absence of gravity and had no trouble giving birth. The principal experiments, however, were on fish — and what's the absence of gravity to a fish?

Still, the birth in Martian orbit may be without mishap except for the political ramifications. Who is the first "Martian"? Is it the first crew member to step out on the soil of the Red Planet, or is it the 8-pound, 3-ounce human squalling its first breath into the anxious faces of the three people aboard the mothership? Her name is Deirdre Channon O'Neil. She has very green eyes. She does not seem bothered by zero gravity. There is a troublesome lack of diapers or their substitute — the spacesuit diaper material works well, but there is a limited supply of it. Also it will be put to use on a tiny human body for which it had not originally been intended.

Even if nothing as exciting as Deirdre's birth happens

on the first expedition to Mars, the media coverage is certain to be a wonderful circus. The crew may become very weary of the inevitable distortions of the sensational press. Suppose crew members report a light green flash on the night of Sol 23. This might become "a Martian beacon seen from a distant ruined temple not far from the landing site," in the future words of a less than respectable, but nevertheless widely read, magazine. In reality, the flash might have been a meteor, high in the atmosphere.

Since the press will be having a grand time with Mars and things Martian, someone will certainly raise the issue of life on Mars. While many biologists feel that living forms originating on Mars are an extremely remote possibility, the subject is of immense fascination to the public. All life forms modify the chemistry of their environment by eating (in some sense of the word) and leaving waste products. Any planet with life will therefore have a "life signature."

Our tired crew will probably find no Martian life forms, since the atmosphere of Mars seems to be completely free of life signatures. So does the soil, if we accept the basic finding of the Viking explorations of 1976. They will certainly not find the wonderful creations of our fiction and our dreams of the planet. Those creatures have gone with the canals, with the great pumping stations, with the gentle race of Martians contemplating the infinite. The hard edge of science will have once again caught up with imaginative fiction.

The explorers will not find that moss covers everything but the poles, as Edgar Rice Burroughs wrote. (That would probably not be news in any event — moss does not capture the public's imagination.) They will not find a Mars that is entirely water covered with continents of seaweed floating on it, as Ludwig Kann described in 1901. They will not find the great Thoat, 10 feet high at the shoulder, four legs on either side, with a broad, flat tail — nor a Martian 15 feet tall astride a beast. They will see none of the cities of Barsoom, either from orbit or upon the ground.

The expedition will not find Weinbaum's crazy leafless plants, the mud cities of the Tweel, or any evidence that Buck Rogers had been there. George O. Smith's Temple of Canalopsis will be absent from the pastel-pink scene. There will be none of that fantastic and wondrous Mars, seen through a glass so slightly darkened, of Ray Bradbury.

Ylla will not lie in her bed to dream of the Earthman. She should have been there, singing some Martian song or wandering by the lush side of a great canal filled with lavendar wine. Of all the Martians the human mind has yet invented, she is the most beautiful, the most human and at the same time the most alien. She is what Mars is all about. But she will not be there . . . or possibly, just possibly, no one will see her. She may be only a creature of the mists, but she is the real Mars, as real as the dirt and rocks in bags that the crew of the lander will bring home to the space station for study by the scientists of Earth. She is real because she is part of the great dream of the Red Planet.

What *is* there is a new world to live on. The half a ton of soil, rock, and other samples — in addition to the landing party's experiences, the tests made by prototype processors, plant growth observations, estimates of resources, and everything else that the expedition will bring back to the space station — will confirm much of what has long been claimed: Mars is not much different from Earth. The measurements of the atmosphere, tests of the permafrost, the water frozen in its original state in safe containers, observations about the weather — all will confirm that Mars can be a second home for us.

Mars is dissimilar because it has traveled down a different evolutionary track than Earth. But it is not so dissimilar that we cannot survive there indefinitely; the similarities are so great that with a few changes and enough time, it could more and more resemble Earth.

Two mobile laboratories will stand parked on a hill in Solis Lacus. The plastic dome of the first expedition's base camp will be empty. Micro-meteorites will have punctured its surface, and it will lie like some great wounded silver manta ray, draped over the blasted crater rim that formed its base. The permafrost solar cooker will sit abandoned, outlined against the pink Martian sky. It will no longer make water.

The electrolysis machine will die; oxygen will no longer flow from it. Here and there about the landscape will be small boxes, large aluminum containers, an abandoned spacesuit. The expedition will leave behind everything it can, but not from any sense of desecration; weight is a terrible penalty aboard a spaceship. It takes fuel to lift weight, even against the gravity of Mars. The crew's first priority will be to get back. They will leave the first junk pile on Mars.

A few hundred yards away from the clutter will be a honeycomb aluminum structure, much larger than anything else in the deserted landscape: the remains of the lander. The ascent stage will have returned to orbit. It will have met and docked with the mothership. Home is many months away, time enough to plan and think, time enough for these few people to ponder what it has meant for them to have been to Mars. They will be the end product of a hundred human dreams, the work of ten thousand engineers. Time enough for Deirdre to fish about in zero-g in an imitation of an infant's crawl with nothing to crawl about on. She will be the most famous human being ever — the most famous Martian.

The Mars ship will not go directly back to the space station. It will take the long orbit halfway around the sun again and dock at a quarantine facility near Earth. While the chance that a Mars organism will exist aboard the ship is low — it is far more probable that the humans and their ship contaminated Mars — the chance must be explored. Even if public opinion is all in favor of the Mars expedition, enthusiasm would not last in the face of even the tiniest rumor of a Martian bacterium. Buried deep within us is the fear of the plague.

The crew will go into quarantine, which will add to their time in space. At least the quarantine station will be much larger than their ship; and best of all, there will be other people around, new faces peering into the viewports at the "Martians." The press will go crazy trying to take photographs with their remote cameras of Deirdre doing anything, anything at all.

Will she go down to Earth in one of the space freighters? A waiting world would like to think so. But she may have spent her entire lifetime in zero gravity. Despite having the bones and sinew of Earth, she will be apart forever. What would she think of the crushing gravity and smelly air of her parents' home planet? Would she have any tiny memory of Mars and of her little spaceship world circling above it?

It will not be long after the first expedition that a second one will be assembled in an orbit near the space station. By the time Deirdre is old enough to go there herself, there will be a city on Mars.

5 BUILD WE MUST!

FROM FOUR HUNDRED miles away, the surface of Deimos, Mars's smaller moon, will look almost smooth, looking superficially like Earth's moon from up close. The astronauts of the second expedition will crowd into the flight deck of their ship to get their first close-up look at their destination.

Deimos has none of the deeply etched linear grooves of Phobos, Mars's other moon, and Deimos has fewer craters. Bright patches of material show up near the intersection of the flat plains of the little moon. As the ship closes in, the surface will become more rugged to the eye. The crew will stare out the forward-facing windows at a savage, boulder-strewn, heavily pitted landscape. Flowing away from many of the craters, the vast majority of them quite small, is a wind-streaking effect that has been caused by dust and debris sucked up by incoming meteors and deposited farther down the meteors' tracks as they ploughed into the surface.

Coming in closer, the people crowding into the flight

deck will see that the surface of Deimos is saturated with small craters and boulders the size of houses. It is a desolate little world, devoid of much interest except that its unique position in the Earth–Mars transfer orbits makes it very important for the future settlement of the Red Planet. There is no life to the alien landscape; it is a dull gray with no discernible color variations. Very little has changed on this moon since the formation of the solar system except that the surface has become more piled up with dust acquired in its millions of years circling Mars.

Slowly the second expedition ship will descend until the speed of approach is only a few feet per second. Below, scattered in dozens of piles separated by a few hundred feet, are bright aluminum containers with staggered red block letters on them. With a final rocket burn, the ship will gently touch the surface. The great globe of Mars will fill the sky, looking very rust-red.

The crew will put on their spacesuits and set foot on deep, powdery dust. They will glide, rather than walk, over to the first group of marked containers. The gravity of Deimos is so slight that a determined run or a fast version of the lunar "kangaroo" would achieve escape velocity. You could jump off the moon if you wanted to.

The two moons of Mars are astronomical oddities. Perhaps not all that odd, since they are much the same as asteroids, which scientists believe they were before being captured by the gravitation of Mars in some primordial time. They are often called the "potato" moons because that's what they most resemble in photographs — diseased potatoes as well. They are unique in the solar system in that one of them, Phobos, goes around its planet three times before the planet has rotated once. Thus the "month" as a lunar reference on Mars occurs three times a day.

Mariner 9 photographed the tiny moons and revealed their heavily cratered surfaces for the first time. In Earth's largest telescopes, they never appeared as more than tiny specks of light about which not much could be deduced except by inference. That they would be relatively heavily cratered was not predicted by any astronomer of note before a spacecraft visited them. The largest crater on Phobos is called Stickney, the maiden name of discoverer Asaph Hall's wife. Hall's discovery came in 1877, the same year the canals were announced by Schiaparelli.

Phobos, the larger moon, has a long diameter of about 14 miles and is 3,700 miles above the Martian surface — about as far as Mariner 4 was when it shot its famous twenty photos. Deimos, with a diameter of about 8 miles, is much farther away: 12,000 miles. It goes around the planet in more than a day — about thirty hours. Both moons rotate, with the rotation of each moon locked into the same period of time that the moon takes to revolve about Mars; thus they always present the same surface toward the planet. Our own moon does the same thing, although it has a slight wobble so that we see around the corner of it from time to time onto the usually hidden side.

The moons of Mars fit right in with the strange pink sky and the rocky, desert landscape. Seen from near the Martian equator, Phobos — the inner moon — rises in the west and a little over four hours later sets in the east. As it goes across the Martian sky, it shows the same phases as our own moon does: new moon, sliver moon, first quarter, half, gibbous, and full. It goes through half this cycle on one pass. It rises every eleven hours or so, seen from the surface.

Deimos is even stranger. Since it revolves around Mars in slightly more than a Martian day, it hangs in the sky for sixty hours until it has exhausted the number of revolutions required for it to appear to "set," as seen from the surface. It too goes through a cycle of phases.

Because of the peculiarities of their orbits around Mars and their tiny size, the two moons can do some other bizarre things. Phobos eclipses the sun 1,300 times a year, but since it is too small to cover the solar disk (as seen from Mars), the eclipse is never what we would call total; it is "annular" and lasts for nineteen seconds. Deimos also eclipses the sun but only 120 times a year; each time the moon covers a tenth of the solar disk for about two minutes. They also occasionally eclipse the sun at the same time, which is surely an event on Mars. The moons not infrequently eclipse each other, too, another weird astronomical event.

Neither moon is large enough or reflective enough to shine much light on the Martian deserts. From the surface of Mars they are brighter than Venus appears in our sky; one could just barely see one's shadow in the light of Phobos if the conditions were right.

It was once proposed that both moons were hollow because of certain peculiarities of their orbital motions that could not be easily explained. This led to a discussion about whether the two moons had once been space stations of a long-vanished Martian race. While the idea was fascinating, it was quickly disproved by spacecraft data. The curious business of the orbital peculiarities was explained by complex — but quite ordinary — science.

Along with all the other oddities about the moons, the Red Planet scores another first in the solar system: Phobos is a "terminal" moon. It is in a delicate gravitational balance with Mars and the balance is degrading over centuries. The orbital decay rate indicates that Phobos will hit the planet's surface in about thirty million years.

Either of the little moons would make an excellent way

station for the colonization of Mars. Most of the scientists and engineers involved in planning a major exploration or colonization of the planet have looked into the possibilities of saving fuel, time, and life support by using the moons as staging areas. They would make a perfect place to establish a supply cache. Deimos would also make an excellent scientific station for studying Mars close up.

The Deimos option, as it is sometimes called, involves sending ships loaded with supplies from the space station to Mars orbit. The ships would then dock at the moon and drop their supplies. This would not require expensive technology or a high fuel expenditure. And it is a safe way to get supplies to Mars orbit because the freight ships need not be manned; they can be robots. Docking maneuvers have been a part of the space program for nearly a quarter of a century.

The supply ships would not use conventional and very expensive fuels. They could use solar sails, which, as the name implies, are a form of sailing using light pressure from the sun. Photons of light fall on a vast sheet of ultra-lightweight reflective material. The plastic film would look like very thin aluminum foil but would feel more like Saran Wrap. The most commonly suggested material is Mylar. Another material is Kapton, which is more resistant to long exposure to heat and is less easily weakened by solar ultraviolet radiation. It is also incredibly light: a solar sail of 3,000 square feet would weigh less than two hundred pounds, so the main weight of a sail craft would be the payload, which is the basic virtue of the concept.

The force applied by photon pressure from the sun is extremely tiny — so tiny that even the thinnest atmosphere would spoil the effect. But in interplanetary space there is no atmosphere. With only the tiny pressure of light, a solar sail and anything attached to it, such as cargo for Deimos and the settlers on Mars, would depart from Earth orbit and travel slowly outward on a long, sweeping path to the Red Planet. The photons' push is constant, so theoretically the eventual speed is that of light itself, because photons by definition travel at light speed. The time needed to get the sail and its cargo up to light speed, however, is measured in years followed by a lot of zeros.

Solar sail cargo ships have numerous advantages, not the least of which is that no crew is needed. The cost of a solar sail craft is low: the sheet plastic, some thin aluminum ribs that can be built in an automatic beam builder, and a computer to run the navigation system are the only requirements. Automatic beam builders have already been designed for use by the space shuttle. They will be used to build giant antennas and other ribbed structures in space.

It has been suggested that one hundred external tanks from the space shuttle could be placed in Earth orbit by 1995. These could easily be retrofitted as cargo carriers, a task for which they might be much more suitable than as remodeled spaceships. The tanks would be attached to a solar sail package and sent to Deimos.

This is a cheap method of building up supplies for colonizing Mars, since the tanks are essentially free and the propulsion — sails — is extremely inexpensive. The cost of sending the supplies to Deimos would, for all practical purposes, be simply the cost of the supplies themselves. The biggest disadvantage to shipping supplies to Deimos by solar sail is that the cargo ships will not get there in a hurry. Nor will they return to Earth rapidly to be refilled at the space station. Thus it will take a long time to establish the Deimos base; but that time can be used to evaluate the scientific data from the first expedition and for building a more suitable, and preferably faster, spaceship for the manned flights that will come after the supplies have arrived.

After the Deimos base has been established, the crews who have been sent there will use all the technology available, working in ways not unlike those of modern scientific expeditions to the Antarctic, to tell us what we need to know about our future on the Red Planet. Orbiters will be released from the little moon; they will take up station over Mars in much the same way the old *Viking* orbiters did in 1976. In the low gravity of Deimos, a shove out of a ship's cargo bay would almost do the job, although the suitability of the resulting orbit might be in doubt. These satellites will map Mars down to a resolution of fifteen feet, report on planet-wide water resources, monitor the global weather patterns, and radar-map the terrain.

Penetrator probes will drop from orbit and burrow deep into the Martian landscape. A torpedo loaded with electronics will drop through the atmosphere, using a small parachute to slow the descent and stabilize the fall. Part of it will impact at extremely high speed, penetrating tens of feet into the soil; the second part will hit the ground rather hard, but not too hard for some kinds of solid-state electronics to survive, and be a broadcast station for seismological data coming from the deeply buried probe.

The Deimos station will also explore Mars with unmanned landers and rovers, looking for the perfect place for a permanent base. It will be the duty of the personnel at Deimos, in a historically permanent way, to make the decision about where the first city on Mars will be.

The manned portions of the Deimos missions will use a modified spaceship powered by ion engines. Ion engines, also called solar-electric engines, will be used for many of the

NASA unmanned robot probes flown between 1995 and 2010. The principle of an ion engine is simple: thrust is provided by using the sun's energy to vaporize fuel, electrically charge it by ionization, and accelerate it to tremendous speeds with high voltage.

An ion rocket flies when the electrical energy is converted into the kinetic energy of the engine exhaust beam. There is no white-hot rocket blast. The engines emit a glowing violet beam of high-energy particles. The gas that is ionized for the exhaust beam, the fuel of the ion rocket, is either a metallic conductor such as mercury or an ionizable fluid like argon. The thrust is very small. It would be pointless to launch such a rocket from Cape Canaveral: the ship would never leave the ground against the fantastic gravity of Earth. But out in empty space, an ion engine can thrust for days upon weeks, upon months.

At the end of the long thrust time, the rocket is going just as fast as a chemical engine could make it go and often faster. Ordinary chemical rocket engines would be used to depart Earth orbit and provide the major thrust toward the Red Planet. They would also be used for in-orbit maneuvering near Mars. The cruise mode of the trip would use ion engines.

The biggest advantage of ion engines is in saving fuel. Not only is a tremendous load of chemical fuel not required, but argon is present in quantity in the atmosphere of Mars. It can be extracted — admittedly with some difficulty — to provide fuel for a return trip. The first Mars base ships, therefore, will use Martian sources of fuel for returning the near-empty ships to Earth, where they will be loaded with more supplies and more colonists.

When it comes time to permanently establish a human colony on the planet, something better than conventional spaceships and ion engines may be necessary. Although astronauts and pioneer sorts who may man a permanent settlement might be trained to make the flight in zero-g, colonists would be in much better shape when they get to Mars if they could have artificial gravity aboard the transport ship. While the first efforts on Mars are in progress, engineers on Earth and aboard the space station will be designing and building the great transport ships so that the colonists can make the trip to the Red Planet in comfortable Martian gravity conditions.

Once the astronauts and scientists of the Deimos base have identified the best location for the first settlement and established the scientific investigation of the whole Martian equation, they will be relieved by a second Deimos expedition. This expedition will be a hybrid group: personnel for the permanent Martian base plus a new group of scientists and engineers to man the scientific and supply station on Deimos. By this point in time supplies will have been arriving on Deimos for several years, even assuming the slow cargo vessels powered by solar sails. The assault on the planet will begin.

Surface personnel numbering about twenty will depart from the Deimos station sometime between 2010 and 2015, using two landers to ferry down to the planet. The initial supplies for establishing the base will also be ferried down. When all of the supplies have arrived, the crew will start building the first city on another world. Unforgiving though the Martian environment may be, we will be able to bring to that world a wide variety of advanced tools, life-support systems, power sources, portable shelters, and the whole gamut of modern technology, just as we have brought them to the polar regions of Earth. With adequate safety precautions, the Martian climate can be endured.

Mars has a wealth of resources. There is water in the

polar caps, bound up in the soil. One estimate puts the northern polar cap at 800 billion tons of ice. Would it not be strange if the first city on Mars — or one of the cities we will later establish — were to build a pipeline from the northern polar cap to bring water to the town? We would indeed be the Martians then — complete with a canal of sorts, one used for much the same purpose the "canals" were used for in Lowell's dream: to bring water to the parched Martian deserts. Lowell was not far off in his priorities, either; water will be one of the most important resources for a permanent base.

But we don't need a pipeline or a gigantic Lowellian irrigation ditch to get water on Mars. Not only do the polar caps have water in the form of ice, but so does the atmosphere, from which water can be extracted. However, the easiest way to get water on Mars is to get it from the permafrost. Using a 50-kilowatt thermal source, perhaps powered by solar energy (or nuclear), somewhere between 60 and 110 tons of water a month could be removed from the Martian soil, even in the drier areas. Water might have been a problem for the mythical Martian race with its fairly large population (always assuming the Martians would have been anything like Earth's people), but it is not much of a problem for an expedition or for a technologically advanced permanent base or colony.

Water is vital for life, but it is the component hydrogen atom of H_2O that is important too. The common denominator in food, plastics, industrial liquids, clothing, fuels, and a host of other items we use daily is the hydrogen atom, the most common atom in the universe. Stars are mostly hydrogen. The Martian water can be robbed of its oxygen for breathing and its hydrogen atoms for making a wide range of useful things.

The Martian air is mainly carbon dioxide, of which a primary part is the two oxygen atoms; water also contains oxygen. We can get breathable air from either of these sources without having to carry tons and tons of it from Earth, as would be the case for a base being established on our moon. Since, however, the Martian soil has plenty of oxygen, it would be the best source of the base's air supply.

There is plenty of sulfur, and there are a thousand useful things Earth people need that have some quantity of sulfur in them: acids, fertilizers, dyes, detergents, DMSO (the neutrino of solvents), steels, explosives, fungicides, and insecticides. On Earth, mankind uses 100 billion pounds of sulfur every year.

There is no presently accepted proof, but the abundance of sulfur and some other elements leads many scientists to think there is phosphorus in the Martian environment too. There are iron, zinc, lead — the list is quite endless. Magnesium chloride can be produced as a salt from the soil, and from that, by fusion electrolysis, comes pure magnesium metal and chlorine gas. Magnesium metal is extremely useful as a structural building item, since the weak Martian gravity favors the use of lightweight metals in construction. Chlorine gas has some interesting uses too, but it is deadly to humans.

Mars seems to have everything, minus organic compounds, that we might need for a permanent base, if we can get at it all. The proportions are a little different from what we are used to on Earth, and everything is in the form of raw ore, or water, or air, or soil; but if we have the required technology, we can make almost anything we have on Earth and use it in much the same way.

The first colony on Mars will need a small town. The first step for the second expedition will be to build it — perhaps before winter sets in. Winter is twice as long on Mars as it is here, although it obviously makes less difference than on Earth since the people will be in environment suits or under domes. Considering the present state of the Martian atmosphere — very thin and thus of little practical help in shielding the surface from radiation and ultraviolet rays — the first city on Mars will be built mostly underground. It may look like a bunch of giant German sausages strung together. Put another way, it may look much like the current ideas for a modular space station, a kind of Lego construction of cylinders.

There will even be some advantages to living on Mars: for one thing, there is no rain, sleet, or snow — regrettably, no postman to defy it, either. (All the mail would be electronically relayed from Earth to the space station, then to Deimos and by microwave down to the Mars base.) There are no floods, mudslides, or ground creep at this point in the history of the planet. Marsquakes are rare and not strong. There aren't any brush or forest fires; there are no hurricanes, tornadoes, or windstorms. Damaging weather events are almost nonexistent; the occasional dust storms are the exception. Natural erosion of buildings would be nil and there would be little if any corrosion of metals. You could safely leave the tractor outside for a few centuries without serious worry.

A minor irritation may be that a compass won't work — at least, we don't think it will. The debate over whether Mars has a sensible magnetic field has been going on now for more than ten years with no definite answer. The best guess is that if Mars has a magnetic field, it is a very weak one and of little use. A simple compass won't work, or won't work very well. In these days of fantastic navigation satellites and sophisticated photo-reconnaissance equipment, the problem will be of little consequence.

If the site chosen for the base is deep in Martian sand, it can be cleared by using portable equipment not unlike snow blowers powered by condensed carbon dioxide. It will also be possible to use any of the usual construction explosives to clear the site. They will work, even in the atmosphere of Mars, for the same reason they work underwater on earth: most explosives have their own oxygen supply in the chemical compounds of the explosive material. Explosives can also be made on Mars from ammonium nitrate, which is nearly as effective as TNT.

Erz cement, a product used extensively in Germany, can be manufactured, although the required calcium sulfate may not be terribly abundant in the Martian soils. Sorel cement can also be made. This material is quite hard and has good strength properties. It is not useful in a humid climate or near water, but that is hardly a problem on Mars.

The cylinder, while not the most perfect form of human dwelling — anyone who has spent much time living in a Quonset hut, which is a half cylinder, can attest to that — can serve very well. A cylinder can be built by putting together easily manufactured sections — quarters, eighths, or even sixteenths. Each cylinder can serve a separate function. One can be a command room or laboratory, another a mess hall and galley, another a garage or sleeping quarters, and so on.

The cylinders could be made of suitably thick Mylar. They could be held in shape by air pressure or by a combination of air pressure and ring sections made of aluminum, titanium, or magnesium. Magnesium is relatively easy to obtain on Mars. Local magnesium products would be lighter than Earth's carbon-fiber plastics, although not as strong.

The interiors of the modules need not be quite as severe as might be imagined. There will be room for furniture, paintings, floor coverings of some kind, and wall hangings. Chairs, tables, and even sofas can be made from magnesium. The use of any plastics is unlikely: plastic making on Mars may be difficult because the organic starting compounds are not present. Plastic objects would be expensive to ship from Earth.

The Martian base could be made at least as comfortable as Antarctic stations on Earth, and for anyone who has visited one, they are not as primitive as we usually think. The environment outside may leave much to be desired, but inside the stations are just about all the comforts of home, including entertainment. Pool tables, however, may not be on the agenda. In addition to the tables' being a bit heavy to cart from Earth and perhaps of a very low priority for manufacturing on Mars, the ancient game of billiards would be a bit tricky to play. The low gravity would probably have the balls flying all over the room unless they were outsized, which would spoil the game, or composed of extremely dense materials (lead, for example), which would be very hard on the cues.

For power, the base could use a small nuclear reactor not unlike that used aboard an atomic submarine. Nuclear submarines have been around for more than thirty years and have been noticeably safe so far as dangers from the reactors go; there is no reason to suppose that a similar reactor could not be used on Mars. Power could also come from a solar-powered satellite. Sunsats, as they are called, have been suggested as power sources for cities on Earth for years.

A sunsat for Mars may be far in the future, however. A nuclear reactor might be a surplus item to Earth's space civilization and could be sent to the Red Planet relatively cheaply. A sunsat might be quite expensive to build and ship and one cannot be easily constructed from native materials. The receiving antenna for the sunsat's microwave power transmissions would also be very expensive to build and establish on the planet.

A greenhouse would be an early priority for the settlement. Experiments in closed ecological systems designed for use in space or for supporting a permanent base on the moon or Mars have not been as successful as we would like, even though the idea has been around since the Russian space travel pioneer Tsiolokovsky mentioned it in 1903. NASA and the U.S. Air Force have been researching farming in space and closed ecologies for more than twenty-five years. But by the time a city on Mars is established, the experiments in closed ecosystems may have long ago provided the answers on how to keep a greenhouse going in an alien environment.

Obviously a greenhouse would not be built just to put nice roses on the table for breakfast, even presuming that roses will grow in modified Martian soil (they should). To grow enough food, the new Martians would have to use what is called "high-density" agriculture, which is now being practiced to some degree on Earth.

The usual Iowa farm sort of agriculture that our grandfathers were used to is not the most efficient. They didn't much care about high production; they had plenty of good land to use. Good farmland on Mars is about as easy to come by as it would be at the South Pole; so some improvements on granddad's methods are needed. The first improvement is a natural consequence of the location: although there is less sunlight, the plants can be grown in an atmosphere rich in carbon dioxide. The efficiency of agriculture on Mars in ordinary circumstances will probably be greater than on Earth.

Special strains can be derived that will produce five times as much yield per acre as we are used to on Earth. There is also the technique of interplanting, which saves growing time by introducing the seeds for a new crop between the plants of the old. The result of interplanting is still more yield per acre.

The crops planted by the colonists will be the same ones we have now: wheat, rice, white potatoes, sweet potatoes, soybeans, peanuts, lettuce, sugar beets, broccoli, corn, peas.

Soybeans and peanuts are especially useful because in addition to being high in protein, they can be processed into oil. Leaf lettuce is a good salad crop and it can be harvested earlier than head lettuce. Strawberries would make good fresh fruit because they grow easily. They also have vitamins B2, B7, and C. Tomatoes would be nice in salads and they are high in potassium and vitamin C. The new Martians will also have carrots, which are high in vitamin A. If nothing else, the carrots may improve their eyesight so they can see the potato moons wheeling about the sky.

With luck and very careful planning, each member of the Martian base could be supplied with a vegetarian diet utilizing an area of only 180 square feet in the greenhouse. For a settlement of one hundred people, the "farm" would be easily within the construction capabilities of the new Martians. In a final edging of the agricultural envelope, it is possible that a sixty-acre farm could support a Martian town of several thousand people for decades. While a farm like this would be hard to establish in the early years of a colonizing effort, it is a realistic goal.

Sooner or later the question of what to do with the leftover plants and vegetable stuff from the greenhouse will come up. In fact, it will come up quite quickly, since human beings are not really herbivorous by nature. We like corn on the cob, but we're not terribly fond of either the cob or the cornstalks. We like the wheat grains and what can be made with them, but we don't have much use for the chaff. Other animals in Earth's ecology do, however; and on Earth we ordinarily eat many of them, since we are fond of a diet high in protein and much of the protein comes from animal meat.

So what to do with the leftover stuff from the garden? Feed it to animals. Carting a load of animals across the void in a kind of Noah's Ark of Space is a touching idea, even a romantic one; but it does not take into account modern biological advances and practices. In the first place, we can forget putting most mammals into the Ark two by two because artificial insemination is widely successful with most species. The billy goat can be left home where he can browse afield as billy goats do and not have to deal with the weak gravity and be called a Martian.

The female is the only one necessary, along with enough frozen sperm from a genetically varied group of males so that the resulting animal batch can reproduce in the normal way without declining due to genetic faults. If a sow has four female piglets by one male and the piglets are inseminated by a different male, the resulting pack, herd, or whatever the Martians decide to call a batch of pigs will be genetically stable, even if partially bred back later on.

The romantic image of the Ark may fall even further behind the times when the Mars colony is built: animals may be carried to Mars as both sperm and ova, to be united only in a genetics laboratory. Test-tube animals.

This possibility is not entirely far-fetched, given that test-tube babies have been with us since 1979 and the experiments leading up to the first human test-tube baby were done with cattle and sheep. Of course, only conception is really achieved in the laboratory; the embryo is implanted in a living host where birth takes place in the normal way. But experiments are already in progress that may eventually allow for a completely artificial birth.

Our Mars base may be built in an era when the shipment of one multipurpose, mammal-duplicating plastic Mother-Thing-in-a-Box is all that will be needed to put a herd of horses on the Red Planet. At the moment, there is not much use for equines on Mars and building a corral would be a waste of time. But it may not always be that way: having horses on Mars may not be as impractical as it seems.

On a more mundane level for the moment, which animals would we introduce to Mars that would make the life of the colonists, if not the animals, better? Perhaps rabbits. They like alfalfa, which with the addition of a bit of salt is quite palatable to them. They, in turn, are reasonably palatable to humans. The meat is low in fatty content and can be cured the same way ham is. It can also be made into sausage or served in a variety of ways including stew.

Rabbits also reproduce rapidly, something for which they are justifiably famous. Thus out of a square yard of Martian soil would come a litter every two months, fed by a few square yards of alfalfa from the greenhouse. The investigators who have worked it all out seem to think that rabbits would provide, on the average, 150 pounds of boneless meat per acre per day. That's more than a beginning Mars colony would ever need, considering the boredom that various forms of cooked and stewed rabbit would engender.

For variety, the garden trash can be fed to goats, which produce milk and cheese. Of course, cows also produce milk; but old bossy might be left at home, since cows produce about four times as much milk as goats while requiring ten times as much feed and weighing ten times as much. We prefer them to goats for meat, however, and it is possible that a cow or two will casually chew a cud or two in the farm on Mars sometime during the early years of the colony.

Chickens can be fed solely on the leftovers from the farm and waste from butchering. The colony can have four or five eggs a week per person without any additional farming area for raising chicken feed. Pigs can also be fed scraps and leafy addenda; the result, in the low gravity, would probably be a very large pig and considerable amounts of pork, lard, and bacon to go with the eggs from the chickens.

Once the town has been established, some of the people will start exploring with highly sophisticated rovers. The rovers will be rather large, self-contained wheeled vehicles able to move over extensive areas of the Martian surface, perhaps able to stay away from the base for days and weeks at a time.

They might be powered by solar panel technology, which converts solar energy to electricity used to power electric motors, but this system may not be efficient enough to get the job done. The big problem with solar energy is that Mars is much farther away from the sun than Earth is.

Internal-combustion engines won't work, since they require air and the Martian air is unusable; it is too thin and so could not be compressed enough. Further, there is little available oxygen to sustain combustion.

The most likely candidate for a rover engine is one powered by hydrogen peroxide motors. Hydrogen peroxide as a fuel has many advantages. It is easily handled as a liquid and it is generally nonhazardous. It can be stored outside the colony and in the Martian environment without harm. It can be obtained by simple electrolysis of aqueous sulfuric acid; the acid can also be produced on Mars.

Hydrogen peroxide has other uses than as fuel. A separate hydrogen peroxide cell aboard the vehicle can power a heating unit as well as provide drinking water and oxygen for breathing. The heat is a by-product of the process that produces water and oxygen. In fact, hydrogen peroxide is about as useful as anything can get. It is H_2O_2, which means two hydrogen atoms and two oxygen — the basics for us. In a pinch, you can even pour some of the fuel on a cut.

Hydrogen peroxide as a rover fuel is certainly more attractive than some of the other substances that have been suggested, one being hot calcium cyanamide, the product of which would be a gas to drive an engine: nitrogen and carbon monoxide. There would be a waste product of calcium oxide trailing the vehicle over the countryside. Another possibility is a clean-burning engine powered by carbon disulfide and the Martian air, but unfortunately both carbon disulfide and the product of the combustion, carbonylsulfide, are highly toxic to the beings of the third planet from the sun.

The rovers will be able to cover a fantastic area of Mars, much more than the older vehicles used by the first expedition. If the final site for the colony is near the first landing area (there are reasons why that might be so, plentiful water being one of them), then the colonists will have a vast panorama to choose from when they begin their travels — most of the Coprates Quadrangle. They will locate Martian resources, gather samples, and perhaps still look for life from time to time.

The limits of the rovers will be reached when they hit extremely rough terrain. There is plenty of that on Mars, and much of it is in the Coprates area. What the colonists may use then is a small airplane for reconnaissance. Airplanes can fly on Mars after a fashion, despite the tenuous air. The first little aircraft may be robot-controlled and will probably be used for low-level photography and for flying into the canyonlands, where orbital photography may not have been too successful.

With an airframe made of carbon-fiber composites, the Martian airplanes would weigh only 80 pounds. The wings would hardly measure 2 inches thick, airfoils with a span of 69 feet and a width of a little over 3 feet. The engine of the aircraft would be a 15-horsepower hydrazine airless type that would drive a conventional 7-foot propeller. The design of the plane is well within the limits of present technology; the hydrazine engine, while it sounds exotic, has already been produced on Earth and has been test-run on the Mini-Sniffer RPV at NASA's Dryden Flight Research Center.

The first planes would have to be carried folded in the supply ships from Earth, but eventually it might be possible to manufacture them on the planet. The Mars aircraft would be a hybrid: while it would use the hydrazine engine and the propeller for level flight, takeoff and landing would be done using a tiny variable-thrust rocket engine.

In addition to aircraft, the colonists have other possibilities for air travel. Balloons can be used, even though it's equivalent to flying one at 95,000 feet on Earth (actually it's closer to 72,000 feet if the lesser gravity of Mars is taken into account). A Mylar gas-bag about 75 feet in diameter containing hydrogen would provide more than a ton of lift. Unfortunately it would require almost all the lift to get it off the ground: the Mylar, rigging, a small engine, and the hydrogen would total almost a ton. The amount left over for cargo would be about 100 pounds, enough for a camera package at least.

Much better than a balloon would be the dirigible airship. The dirigible, or rigid, airship differs from a balloon in many ways, the principal one being that there is a structural framework inside the gas-bag. While the *Hindenburg* disaster remains the worst example of what can happen to one (the hydrogen gas-bag of the dirigible caught fire and exploded, which is why helium is now used), it is often forgotten that millions of safe miles were covered by the dirigible airship between 1915 and 1930.

A dirigible could use hydrogen produced at the base. What happened to the *Hindenburg* is not possible on Mars: hydrogen won't burn in the oxygen-poor air. With a combination of hydrogen buoyancy and a gas-bag shaped so that it acts as a wing, a Martian dirigible would be very practical. A small one — 100 feet long — could carry 300 pounds of cargo and have a range of more than 100 miles. Large dirigibles could be built to carry passengers.

Balloons, airplanes, and dirigibles could all be used for remote-control photography of the Martian surface from low altitude. Controllers at the Martian base could send the craft almost anywhere they wished within the cruising radius. The photography would be unparalleled in human experience, and if the Martians wanted to, they could produce an exploration series for weekly television broadcast that would be more fascinating than any ever seen before.

Looking straight down into the summit caldera of the gigantic volcano Olympus Mons, a camera could photograph the 15-mile-wide floor. The floor is covered with small impact craters and irregular volcanic vents. Panning down the surrounding slope of the volcano, the camera would record irregular channels and collapsed lava tubes as well as the strange "wrinkle ridges" first identified in photographs taken by *Mariner 9*.

Scientists on Mars could use the robot exploration from the air to photograph the southernmost regions of the planet: frosty Hellas Basin, the strange eyeball-like feature that is the crater Schroeter, or the dark, sinuous terrain at the western edge of Syrtis Major, the wind-streaked plain that may be a secondary landing site for manned missions to Mars. A third of the way around the planet to the east from Syrtis Major, the cameras could record the jagged rim of the Gustav crater and follow the large channel Ma'adim Vallis that winds southward from the crater for nearly 700 miles like the tail of some gigantic tadpole.

For the geologists on Mars (but would they be called

geo-logists on the Red Planet?) the aerial reconnaissance would be the key to solving some of the most fascinating and puzzling questions about the planet. They could explore the geological nightmare of 300-mile-long Mangala Valles. Mangala Valles has everything but the geological kitchen sink: it is a tremendous water-carved channel that also shows evidence of having lava flows, meteor impacts, windborne dust deposits, wind erosion, and possibly "mass-wasting" (the Martian equivalent of a California mudslide).

If the aerial reconnaissance craft could be equipped with robot shovel arms, they could eventually be directed to land and take samples from over the whole surface of the planet. One controversy the wide sampling might resolve is why Mars is red. It has generally been concluded that the red soil of Mars is a form of oxidized iron (the principal candidate being maghenmite). But it has also been suggested that the soil is heavily composed of feroxyhyte. Either way, the soil is rusty; that's what makes it red. But whether the same basic dirt exists over the entire planet is something we may not know until scientific expeditions can roam over the red world and test the soils from a thousand locations.

6 THE CANTERBURY PILGRIMS

BY WHAT GREAT WITCHERY can we foresee the events taking place in the first city on Mars in the early years of the twenty-first century? A glance at what passed for everyday life in the time of King Arthur will show that a thousand years brought forth a giant leap for this tribe of intelligent animals called the human race. That leap was made at a snail's pace compared with the changes that have come about in the last 500 years. Can we really peer half a century ahead and predict with any accuracy what might transpire?

Imagine a man of the seventeenth century trying to puzzle out a day in the Europe of 1990, after being suddenly and inexplicably transported there. He would be familiar with city streets and roads; houses would not be terribly different. But what would he make of the telephone? Would not a microwave oven seem run by magic spirits?

Could he explain the contrail of a high-flying jet, or would he cower in the street and avert his eyes from what was clearly the devil's work? If he survived the initial shock, he might adapt reasonably well, despite the language barrier; but truly nothing in his medieval world-view would prepare him even remotely for life in his descendants' future.

Having seen the incredible advances in technology since 1945, we are in a much better position than our mythical seventeenth-century time traveler. We have some experience upon which to base our gazing into the crystal ball. In this we are not unlike our cousins the Edwardians. They lived in an age blazing forth with new ideas and new inventions. The steam railroads made a shambles of the notion that people lived and died a few miles from their birthplaces. Marconi had perfected the radio; the telephone was in commercial use. The telegraph had long been a part of their world. The airplane was flying — little bare carriages of light wood, dope, and linen.

Men drove their grooms and footmen crazy with manuals and advice on taking care of that new invention, the automobile; the horses were left behind to drowse in the field, to stare blankly at the puffing, clanging contraptions that had taken over the roads. The Edwardians were standing on top of the whole future, the looking glass of the new century; many of them saw it clearly, and even the least imaginative could see with some clarity what was likely to come.

We live in a similar time. The space shuttle has been perfected and it lifts off from Cape Canaveral so often now that we do not even pay attention when news of it comes instantly to the electronic entertainment boxes in our living room or the other rooms of the house. We will have sent spacecraft to eight of the nine known planets by 1989. We have landed spacecraft on three bodies in the solar system. In a few years we'll make that four with a robot landing on Titan, the giant moon of Saturn, a moon so large that it is bigger than Mercury; a moon with an atmosphere that we now think of as primordial and frozen in time since the early days of the solar system.

We are already using space in a new industrial revolution: the communications satellites, the weather satellites, artificial moons for navigation — and eventually, no doubt, war. We are building a space station where men and women will work and spend their leisure time semipermanently high above the blue and green world where they were born. Looking only a little way into the future, we can predict factories in space, perhaps waste dumps (instead of polluting Earth forever). We can see bases and mines on the moon. All the technology of the twenty-first century is before us in its elementary forms; it is within our grasp, even though some of it is still on the drawing boards.

Is it any wonder we can gaze dimly into our children's future and see men and women alive and at work in the towns of Mars? While we may not see them as clearly as we would like, we have been prepared by our own past for this vision. We can see the outline, shimmering amid the Martian sands, of a small city peopled by soft-focus glimpses of ourselves. We can imagine them with their hopes and dreams, going about their daily business.

The air they breathe will be recycled many times. No one knows yet if the breathing of recycled air for decades will be harmful to humans. But they will certainly breathe much less oxygen on Mars, and since one of the limits to longevity on Earth for all air-breathing animals is our life-giving oxygen (it is slightly carcinogenic), they will probably live longer than we do. Over a lifetime, there is sufficient danger in our oxygen intake on Earth that we come statistically closer to cancer from it.

They will live by strict environmental rules: waste no

water. That most precious of substances to us humans will not be easily replaced on Mars. While there are expected to be large deposits of subsurface ice from which water can be extracted, it is a difficult extraction. Expeditions to Earth's Arctic regions long ago discovered that permafrost is famous for destroying drill bits, which are expensive and hard to replace at the distance of Mars.

The water from Martian permafrost may not be particularly pure, so more — and expensive — technology will be needed to make it usable for drinking. The Martian subsurface waters may be more like Dead Sea water, which is a bit like a "soup" that is full of dissolved salts. They would have to be removed.

The other Martian source for water, the atmosphere, is an even more difficult proposition. While a colony can be supplied with water extracted from the air, the water will be doubly precious. There is about 0.03 percent water vapor in the Martian atmosphere, depending on the time of year and the area; turning that into water for our use will be hard.

There will be practically no Earth water on Mars, ever. The expedition ships may have some supplies, but those will be generated aboard ship from liquid hydrogen and liquid oxygen fuels. It is presumed that any "holy water" in use on Mars would be of the homegrown variety, blessed on the spot. But as an illustration of the hopelessness of carrying any water from Earth, a one-ounce vial of holy water from our planet taken to Mars would cost the colonists about $10,000.

Air to breathe may never be in short supply on Mars — that would mean the end of the colony — but it will be an expensive thing to produce. Conservation of air will be a primary environmental rule for the people on the Red Planet, just as it will be for a lunar base. Airlocks to the outside will be used as infrequently as possible; access to environment suits will be limited; not everybody will be able to walk around on the Martian sands when they want to.

The colonists will be a community in the old-fashioned sense, in the way that a small platoon of hand-picked soldiers is closely knit, each dependent on both the professionalism and the caution of the other. The base will be organized chaos, a bustling, sprawling frontier town of more and more modules and tunnels. Life there will be fragile; survival will depend on unremitting vigilance with machinery and with safety rules.

Despite the precautions, there will be deaths at the colony. Someone will be careless in an airlock; someone may wander too far into the desert with too little oxygen. Environment suits may be insufficiently checked out just one time too often and the result will be death, sudden and relentlessly awful. Death will be the unseen presence, the shadow figure behind human actions and reactions. It is because existence on Mars will be so tentative at first that the selection of personnel will be so carefully done.

It is characteristic of human exploration that the first ones to arrive in a hostile and undeveloped environment are the hardiest or sometimes the most desperate of our race. They have been driven people, people who could thrust their own mean spirit or indomitable will against savage conditions and win — or at least force Nature to a draw.

The first astronauts were all test pilots. They were trained beyond all eventuality; they could survive almost anything. They pioneered the conquest of space. Now the rules are beginning to relax. Passengers with a few months' training will board the space shuttle. Soon travel agents will book suborbital flights for anyone with enough cash who can also pass a rudimentary physical. It is a pattern we have set for centuries: explorers have always been eventually replaced by bankers, lawyers, hoteliers, and people across a whole range of occupations.

By the time the first settlement ships leave Earth's orbit for the long trip to the colony on Mars, the people who will board those ships will be at least as ordinary as those who ultimately went west in covered wagons in the last century. They will be the kind of people who like settling and living in a harsh, new environment. Obviously packing up and leaving all that is familiar to pioneer a new world won't be for everyone; otherwise the cities of the East would have emptied between 1830 and 1890 as the American West was settled.

Will the colonists respect Mars? Or will they be a juggernaut of technology, tearing and carving up the landscape? Perhaps a bit of both. While it is most likely true that Mars has never had life in any form — and our life dates back perhaps 3.8 billion years — it does not mean there are no ghosts on the Red World. We have been dreaming, writing, planning, and fantasizing about this little piece of the solar system for hundreds of years. We have given it life many times and in many forms. Surely some of the human energy that we have poured into the planet has infected us in some way. When we go there to live, might we not *feel* something about the place, some vague twinge of uneasiness, some sense of the presence of the "old ones," even though they have never existed in our science?

We are a strange race, we humans. Beneath all that knowledge we have acquired is a pagan memory. Our conscious minds tell us that superstitions are false gods, and yet we sense in the dim mists of random thought that there is something out there just beyond our ken. For all our science, can we stay the night in the fairy hills and say in the morning with a cold eye that nothing transpired except the heavens wheeled above as they always do and finally the sun rose? Can we live on that vast and wonderful world fourth from the sun and not reinvent the "old ones" of Mars?

In the modules and cubicles of the colony, the people will live out their lives in ways not terribly unlike our own. They will marry, although they may do so in serial fashion: once for love, once for children, and once for companionship. Human serial marriage for different reasons at different times in our lives was first predicted by Margaret Mead many years ago.

They will be reasonably healthy by the standards of the twenty-first century, very healthy by our standards. The lesser gravity of Mars will produce less strain on internal organs, especially the heart. Asthma, high blood pressure, angina, and motion sickness will all be taken care of through the use of a small skin patch. For each problem the patch will be permeated with a specific drug, and the dosage will creep through the skin in tiny amounts in a continuous buildup.

Blindness and deafness will be relatively unknown to the colonists, partially because medical science will have replaced the scalpel with the laser, paving the way for new forms of microsurgery — already being performed in many hospitals. The ordinary subconscious fears with which we are universally equipped on Earth may be dealt with by a host of new drugs. Some of the drugs will be developed in laboratories aboard the space station between 1992 and 2000. Drug experiments have already been done aboard the space shuttle, with varying degrees of success.

Since the colony on Mars will be quite small and crowded at first, the colonists may use drugs to control the fear of crowds and the fear of closed-in spaces. If the colonists want it, George Orwell's "pleasure pill" will be available: new drugs developed in zero-g may make Lithium and Tofranil look like ancient elixirs from an alchemist. The new mood-altering drugs will be without the side effects we now associate with such things, including the rather nasty side effect of addiction.

A colonist with a gallstone may have it dissolved with drugs or sound waves instead of removed by surgery. Duodenal ulcers will also be treated by pills alone. Hysterectomies will be rare, if done at all, hormonal treatment being preferred. An inflamed appendix will be treated with superpowerful antibiotics.

On Mars in the next century, a bad burn will be treated with drugs and then artificial skin grafted on. Artificial blood made from chemicals called perfluorocarbons will be used regularly. It is even possible that cloned body parts may be available to colonists injured in accidents. While cloned body parts (natural replacement parts obtained by cloning from the original human's cells) are still far in our future looked at from *this* side of the year 2000, they may not always be so.

But the single thing that may lead to long lives for the Martians may be simple diet. We have learned much about aging and diet in the past few decades and expect to learn more in the future. Diets in the colony will be personalized by computer. As a rule, a healthy human body requires about forty different chemicals, but no two people have quite the same requirements. Computers will analyze blood samples and chart how the body handles the food intake. The personalized diets will be varied depending on the type and amount of work being done.

Not only will computers monitor diets, but they will monitor diseases as well. The diagnostic computer has already been developed to some extent. By the time colonies are established on Mars, the diagnostic computer may have replaced the human doctor for that specific purpose.

An important part of colony life will be exercise so that

the muscles of Earth won't atrophy in the lesser gravity of Mars. "Pumping iron" would require a lot more iron be pumped because the exercise involved is much less in the reduced gravity. The barbell would actually weigh less on Mars (500 pounds would seem like 190 pounds on Earth), but it would be just as awkward to handle as you would expect and nearly as dangerous. Other exercise devices we normally use on Earth would also not work too well on the Red Planet, primarily because of the lesser gravity.

Many Earth sports could be played but would require adaptation to the Martian conditions. A normal baseball could be slammed two to three times as far as on Earth and thus would probably puncture the dome. The consequences would be disastrous unless the dome had a quick-sealing compound that would automatically rush to the site of a puncture. Of course, you could substitute a heavier ball, but then the bats (which would be expensive to bring from Earth) would break. Heavier bats could be made, but would the game still be baseball?

Much the same problem holds true for other Earth games: a football pass would easily go farther than a hundred yards. But since a 100-yard-long dome for sports would be a definite luxury for a colony, football may be out for a time. The lesser Martian gravity also makes playing basketball difficult unless the court were much larger than we are familiar with. Shot-putting is also out, and so for obvious reasons is javelin throwing.

The most easily adapted games for Martian conditions are handball, variations on racketball, and table tennis. A combination of heavier balls and larger playing areas would make them as popular on Mars as they are on Earth; working out the game in the new environment, however, would be an interesting challenge.

One sport that will be out for a long time is swimming. A pool on Mars would certainly be easy enough to construct, but filling it with water — and maintaining purity — might be prohibitively expensive for the colonists.

If a horse ever got to Mars, it would be able to outdo anything *Ars Longa* was ever supposed to have done. A horse could jump over an obstacle at least 15 feet high and possibly a good deal higher. Badly designed as they are, horses would be quite comfortable on Mars since the lesser gravity would take some of the weight off their feet and legs. Equine legs, in cross-sectional area, are of insufficient size to support the horse's weight on Earth, which is why horses have so many leg problems and aren't raced much past the age of four.

Our colonists will watch old movies on tape, play stereo sets (taped and live broadcasts), read books, and do other forms of sedate entertainment humans have devised over the centuries. Card games are easy on Mars, but the cards will need to be of a heavier weight cardboard. Chess, checkers, and any other board games are fine if the pieces are heavier.

The original permanent Mars base will be expanded many times in the first decade of settlement. At first the colony will look like the base of the first expedition: a couple of semi-transparent domes, a cluster of generating plants, and a small farm. Later, more domes will be added, along with interconnecting tunnels. Communication antennas, giant white dishes, will dot the red landscape. An open dome with an airlock into the community will house the rovers for outland sorties.

As the settlement expands, more and more native materials will be used for construction. The original thin plastic domes will be replaced with massive, concrete-like modules buried in the soil; except for the lack of provision for planetquakes, the Martian colony may resemble the tunnels and modules of the North American Air Defense Command under

Viking Orbiter 2 approaches the dawn side of Mars. Toward the left, the giant volcano Ascraeus Mons, with a plume of water-ice cloud. In the center, the Martian "Grand Canyon" of Valles Marineris. Right, the Argyre Basin, whitened with frost. (National Aeronautics and Space Administration)

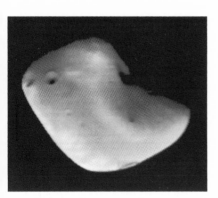

Deimos, Mars's smaller moon. (JPL Image Processing Laboratory)

Grooves on the surface of Phobos, revealed by *Viking Orbiter 1* from 300 kilometers away. (National Aeronautics and Space Administration)

Desert landscape and sand dunes, photographed by *Viking Lander 1* at Chryse Planitia. (National Aeronautics and Space Administration)

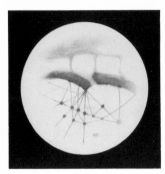

Left: The frontispiece of Lowell's book showed his view of the "canals" from the Lowell Observatory in Flagstaff, Arizona. (Percival Lowell, *Mars*, Boston: Houghton Mifflin Company, 1895)

Below: Water may once have flowed through this channel northward (lower left to upper right) as photographed by *Mariner 9*. (Jet Propulsion Laboratory)

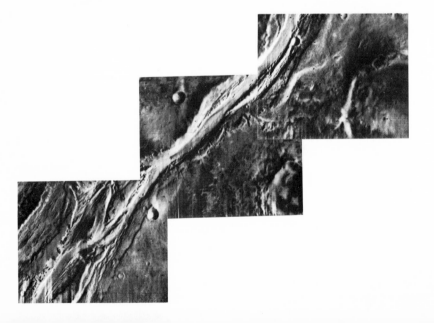

Left: Martian canyonlands: Labyrinthus Noctis at left, Tithonium Chasma at top. A *Viking Orbiter 1* photo. (National Aeronautics and Space Administration)

Below: A portion of the north polar ice cap, photographed by *Viking Orbiter 2.* (National Aeronautics and Space Administration)

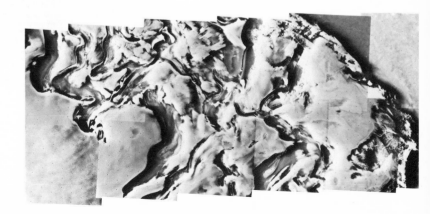

Tests of the *Viking* lander in a space simulation chamber at Martin Marietta. (Courtesy of Martin Marietta Denver Aerospace)

A Space Operations Center designed to stay in orbit some 200–250 miles above Earth. Created by Boeing Aerospace Company and painted by Jack Olson. (Courtesy of Boeing Aerospace Company)

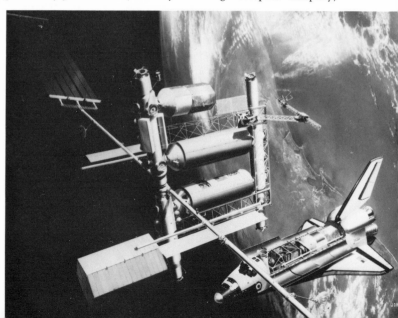

Cheyenne Mountain near Colorado Springs. The purpose is not entirely dissimilar — survival in an environment expected to be harsh — and the construction may not be all that different.

Eventually the city may resemble a giant starfish, with an outer circle touching the tips of the "arms." The arms will be long, wide thoroughfares reaching outward from a central hub; tunnels would radiate from the arms on both sides. One arm might be living quarters for single men and women; an opposite arm might be for couples and married colonists. Another might house the scientific laboratories with an airlock and outside dome for rovers and airplanes. Another arm might house the administrative and colonial offices — even an office for tourism.

Tourism on Mars may not be a dim and future event; when it becomes remotely practical to do so, some enterprising travel agent will book a trip there. In our history, tourism has always followed exploration sooner or later. Shackleton's Imperial Trans-Antarctic Expedition of 1914 was swallowed up in the wastes of the Weddell Sea, its fate unknown until almost two years later. That the expedition survived at all was a miracle: after nearly a year on the ice, and an incredible voyage through the Drake Passage in an open boat, Shackleton stumbled onto the streets of South Georgia to get help for his marooned men.

Today, however, half a dozen tourist agencies will book a trip to Antarctica, usually to the Argentine or Chilean scientific stations. There is even a hotel in the Antarctic with a small staff. You can climb into your car after basking in the desert heat of Palm Springs, drive to Los Angeles International, and several minutes later be bound for the South Pole. All in a big tour package. So it is not surprising that Mars will see tourists, and perhaps not long after the colony has been established. The biggest drawback for hardened adventure-travelers, aside from what would inevitably be a monumental American Express bill, is that the package would be for at least two years.

In the hub of Mars's first city, there may be a miniature of the entire design, a micro-starfish called Lowell Park. In the center of the park would be a statue of old Percival, seated, as if he were at the telescope staring at those intricate "steel engraving" canals he imagined he saw. He once wrote that the telescope was just a form of travel, that when he was seated at the eyepiece on a cold winter night with the dome open to the stars and one red "star" in particular in his eyepiece, he felt that he had embarked upon a journey of discovery, a voyage as real and breathtaking as any on Earth. In a way he was right. His traveling by telescope sparked the human dream that will make the Mars colony; it is only fitting that he should be there in native magnesium effigy to see it all in spirit.

The starfish arms of Lowell Park will be pathways through the grass, the only grass growing on Mars. Here the children born on the Red Planet can see the green (or perhaps it will be Kentucky bluegrass) carpet that grows so easily on their parents' home planet and takes the volunteer efforts of fifty colonists working hundreds of hours each quarter to tend and keep healthy. Around the starfish arms of the park's pathways will be concentric circles, mirrors in miniature of the ring road that circles the city. The innermost circle will be called Bradbury Walk.

The hub of Mars City will have shops. Not unlike the shopping district of any frontier community, they will feature some imported goods (at outrageous prices) and local products, the local products being merchandise from the farm, magnesium furniture, spun metallic clothing, and a few other odds and ends that might be available during the first ten or twenty years of the colony. In such a small community, money as we know it is unlikely; it is far more possible that the entire town will be run along the lines of the classic company store. The old Hudson's Bay Company in the early days of Canadian exploration and settlement provides a good example, as well as a bad one, of how that form of commerce can work.

One of the most valuable things on Mars would be something made of wood. Mars has no trees, of course, and the colony is certainly not going to be in a position to grow them for tables and chairs. Almost as valuable as anything made from wood might be a plastic object, since only the most complex of manufacturing processes could produce plastics on Mars, where the essential elements are not available. The most valuable item on the planet may be somebody's Queen Anne chair, if anyone is ever rich enough to send one there. A native magnesium alloy chair might cost the equivalent of two days' work for a colonist. Shipping a genuine antique chair from Earth to Mars would cost the equivalent of a year's pay for a dozen colonists.

Tempting though it is to theorize the politics of a populated Mars, it is safer to say that the planet, at least in the first fifty years, is unlikely to have a political system very different from those found on Earth. The company store may be replaced someday by a mayor and a town council, but otherwise we would be quite familiar, perhaps too much so, with the antics of the ambitious on the Red Planet.

Our colonists will be in constant communication with Earth through a series of stationary satellites in Mars orbit. Like our own stationary communications satellites, they will allow messages and information to be sent and received at any time of the day or night — with the usual average time lag of twenty minutes to Earth that the signal takes.

The people on Mars will be on their own in a harsh environment; real help will be more than six months away. They will become self-sufficient and independent as rapidly as possible, not from any political intrigue but simply because the situation calls for autonomy. Trying to enforce Earth's ideas on Mars would be a little like trying to discipline a fish: tapping on the aquarium's glass doesn't do much good.

As in most human colonies, there will be sufficient friction eventually that a significant number of colonists may want to go off on their own and set up something more to their liking. Humans have always branched out into new beachheads and formed new communities along different lines. It is why New Orleans differs from San Francisco or Santa Fe or Philadelphia, despite superficial similarities.

Some of the Martians may venture into the volcanic wilderness of far Elysium or park their rovers on the edges of the great canyons of Noctis Labyrinthus. They may stand in their environment suits on the rippled edges of the fading polar caps. They may even journey to the curious "face" of Mars. There is a strange mesa that looks (at least from orbital altitudes) like a human head, and it is next to a five-sided pyramid among other pyramids that some people have thought were related in a mathematical way. With a few million eroded mesas on the Red Planet, it is not surprising that one vaguely resembles something recognizable in the same way that clouds assume shapes, although we know they are natural formations. Still, it is another Martian mystery, and our colonists will eventually go there to see what it really is.

Along with the wonderful exploring they will do, the colonists will always wait with anticipation for the "shuttles" to come in from Earth. Each arrival will be a major event in the life of the colony. They will be called shuttles because that is a convenient description, but they will be ungainly things festooned with antennas, cargo bays, and bulging electronic pods. There is no reason for them to be streamlined: they never enter an atmosphere.

The shuttles will travel in great interplanetary loops, six months to Mars, eighteen months home. The new colonists aboard them will travel in the relative comfort of Martian-equivalent gravity, because the shuttles will spin to provide it. They will dock at the station on Deimos and unload their cargoes of supplies, materials, and prospective Martians.

The colony will become less and less dependent on Earth as each year passes. By the time it has been firmly established with a population of three hundred, the major part of the fuel needed for shuttle returns to Earth's space station will be manufactured on Mars from atmospheric argon. The colony will also become rapidly self-sufficient in food production; the colonists will manufacture most of their own building materials, including structural beams made out of native magnesium.

The colonists will, by this time, probably think of themselves as Martians, not simply as transplanted Earth people. Their children, who will have been born under the strange pink sky on the dusty plain of Solis Planum, will be marrying within a decade or at the most two, and having children of their own: second-generation Martians. One of those who will be marrying is a girl named Deirdre Channon O'Neil.

By the time Mars City is firmly established, she will be twenty-eight years old, and weak gravity will be all she has ever known. She will settle, perhaps, in Solis Lacus, where the first expedition landed and her father first stepped out onto the jumbled landscape of Mars. It will be as much a home as she has ever had because all her homes will have been alien by our standards. Deirdre will have been the first child of space, or at least the first interplanetary child, and she will have been raised aboard a ship bound for Earth from Mars and aboard a metal kindergarten called a space station.

Perhaps a heart murmur forever locked her into the prison of the space station. Perhaps the experience of being born in zero gravity was, in some strange way, congenitally harmful to her. She may never have set foot on Earth. She may carry Mars in her blood, and when the colony ships began departing from the orbit of the space station she may have been one of the first to volunteer for duty at the Mars base. The weak Martian gravity will feel strange to her, if she has spent all her life in zero-g, but she will adapt easily. The human race is marvelously adaptable. She will find the air in the settlement different from that in the space station, but it will not be so different that she will notice it after a few days.

She will live and work on Mars as the others will: hard, difficult work in a frontier community where leisure time at first will be unusual. She will have more status there than in the glare of Earth's press, where she was simply an oddity to be resurrected each year in a soap opera story when editors were looking around for something to fill the pages on the anniversary of the first expedition to Mars. After all, in Solis Lacus, she is the first Martian, Deirdre the Original. Perhaps she will become the mayor on some bright day in the future.

Imagine her day on Mars. She has a room in the single women's dormitory. Her roommate is a reconnaissance pilot, a skilled technician in the tricky world of flying an aircraft in the thin air of the planet. Their dormitory cubicle is only 10 feet by 10 feet. The walls are Martian cement, sealed with chemicals to keep the $-100°F$ cold of the soil from penetrating the room. The floor is covered with a thin sheet of magnesium alloy, impregnated with a pastel luster. It almost glows in the dark.

Their bunks fold out from the wall at night and fold

into it by day so they can use the rest of the tiny floor space. Against a wall where a window would be on Earth is a painting of Mars done by one of the second expedition's crew. It was given to Deirdre at the space station when the second expedition came back from Mars. She was allowed to bring it as personal baggage when she immigrated. On another wall is a spun magnesium tapestry done by a local artist; the colors of the tiny spun fibers catch the light and reflect it in a myriad of shades, depending on the light. There is an alloy chair, a metal desk with metal drawers.

Deirdre gets up at 7:30 A.M. Martian local time. She is a geologist like her father. Her roommate has spent the last week on an outland expedition and is off duty.

Deirdre dresses in her working uniform, not unlike others in the colony: a spun tunic, resembling more than anything else a tunic of the Bronze Age in England, the kind that would have been fastened by a hammered broach in the ancient days of Earth's history. On Mars the tunic is held by an alloy repoussé pin. In the odd way the Martians have adopted whatever pleases them at the moment, the repoussé figure on the broach Deirdre is wearing is Siva Nataraja, the Hindu Lord of the Dance. The tunic top is complimented with a pair of spun metallic harem pajamas. Her shoes are high-buttoned metallic boots with synthetic soles.

Deirdre will leave her cubicle at 7:45 A.M. and walk down the tunnel from the single women's dormitory to the wide thoroughfare that is one of the "arms" of the city. In the center of the street are the thin metal rails of the subway that lead to the hub of the town and Lowell Park. Only the main living areas are served by the subway as yet; two of the city's arms have the little magnesium alloy trains. The subways also run around the city on the ring road. Deirdre works in the scientific laboratories where the geologists, meteorologists, and exploration teams ply their trades. It is quicker for her to ride the subway to the hub and then walk the rest of the way to the labs than to use the ring train.

She catches the subway, a seven-car "uptown" train. Each car has four alloy seats. The cars are open, since they always run underground. The Martians have a distinctive flair for directions in their colony: "uptown" is toward the hub, "out of town" is away from the hub along one of the arms, "around town" is a location along the ring road. Thus to get to Deirdre's laboratory, in the most distant outland section where the rovers and airplanes are housed, one goes uptown to the hub, out of town on radial 4 to module 65 — that is, of course, the direction from the women's dormitory. It is also around town at the terminus of radial 4.

Deirdre gets off the subway at the circular track surrounding the hub. The train continues along the circle and darts down the arm that goes to the married colonists' quarters. She walks along Bradbury Walk until she is opposite radial 4, the arm of the city that leads ultimately to a dome with an airlock. It is a half-mile walk to the geology labs and the outland dome. Today Deirdre will examine the specimens picked up from Outland Sortie #0456721, a serial number designating the Martian date and the particular robot rover that located the samples. The First Lady of Mars will spend a day puzzling over the rocks of what is her native land.

To most of the colonists, especially to Deirdre, Earth will be a distant and increasingly strange place. Although the people of the two planets will be in constant touch through television, video films, and in a hundred more subtle ways, the customs of the home planet will be different, somehow out of place and step with a raw, hardy frontier culture where people are fantastically interdependent for survival and progress. Earth will seem more progressive, too liberal. It is a quirk of human history that a colony is a rudimentary time machine.

The colonists will probably retain many of the customs and values that were popular and accepted on Earth when they *left*. Change will come slowly; the ceaseless struggle to create a new community and a new environment will take up so much of the colonists' lives that they will remain in some ways socially frozen in time.

This is not unusual: there were pockets of people in America who spoke a reasonable brand of Elizabethan English well into the nineteenth century. The people of Easter Island landed there in some forgotten epoch and remained an isolated culture for almost a thousand years. When Europeans found them, they would have been difficult to distinguish from their ancient ancestors.

The settlers will have ceaselessly explored Mars from the surface, using rovers and robot airplanes. They will have identified major mineral deposits and resources, especially areas of thick permafrost. They will expand the base into several suburbs situated around the original hub. The farm will be flourishing under a large dome partly buried in the regolith — the topsoil — of the planet. The colonizing of Mars will have succeeded, with the only major drawback being the restriction of living underground.

One of the Martians' most interesting scientific duties may be in one part of the community they will call Setiville, derived from SETI, the Search for Extraterrestrial Intelligence. We have been listening for decades to various transmission bands in the hope of "hearing" from some other civilization out there and have been broadcasting messages in case someone out there might be listening.

In 1983 mankind launched a program to search the universe for radio signals using a spectrum analyzer. It could monitor 70,000 separate microwave channels and test them for signs of intelligent signals. Future SETI programs have been designed — and may be in place long before we land on Mars — that can examine and analyze 10 million channels in the microwave spectrum. The theory has been that if there is any civilization out there, it must be leaving a trace somewhere in the vast radio spectrum. We are certainly doing so and have been for years with our radio, television, and other transmissions.

The best false signal, one that had astronomers going for a few days, later turned out to be a supersecret radar emission from a Department of Defense airplane. Angels, seraphim, cherubim, and the archangels Michael and Gabriel evidently aren't bothering to broadcast in microwave frequencies. Neither is anything else.

As it happens, Mars may become the center for SETI investigation — or at least the major broadcasting station for messages beamed to the galaxy. There is a natural laser on Mars that operates full-time on the day side of the planet. It is generated because molecules of carbon dioxide in the atmosphere are stimulated into enhanced emissions by the photons from the sun. The natural laser's output is extremely diffuse — the energy is spread all over the sunlit side of the planet. It is not the only natural laser found in the solar system; there is believed to be one operating in the dense atmosphere of Venus. It was, however, a bit unexpected on Mars. When it was first discovered, it was quickly pointed out by scientists that it was not the product of a Martian race, merely a natural event.

Like all lasers, this natural one can be concentrated. If the scientists of Earth decide to use the laser, they might build two huge mirrors in synchronous orbit over Mars. The mirrors would face each other so the path between them passed through the atmosphere at an altitude of about 45 miles. The mirror path would be at 90 degrees to the incoming photons from the sun. The net effect of the mirrors would be to create what is called a "laser cavity," and the energy of the natural laser would be increased each time it bounced from mirror to mirror.

If a system of mirrors could be used, as large as 6 or 7 miles in diameter, the power output would be staggering. A laser shining with that much intensity from the Martian atmosphere in only the CO_2 wavelengths would be vastly brighter than our sun is in the same wavelengths. It would be very noticeable to anyone looking in this direction. If the laser were modulated — a flashing laser beam — it could be seen in a large telescope as a one-second burst from the other side of the galaxy.

Since the laser is a natural phenomenon of Mars, the colonists would be responsible for maintaining the cosmic beacon and the pulsing rate. The laser would be useless for anything else. If it were aimed at Earth — perhaps in a serious gesture of throwing off the colonial yoke — it would not make a good weapon. Upon striking the surface of Earth, the beam would sweep an area more than 7 miles wide, but the energy density wouldn't be enough to make your hand warm.

Who owns the Martian base may one day become a thorny question. The Outer Space Treaty of 1967 clearly says that the spacecraft and "objects landed or constructed" on a planet are owned by the nation that launches the expedition. This seems clear enough, but it really isn't. A space lawyer might point out that this only implies things constructed from Earth materials. The treaty says nothing about objects built of native materials.

It does appear, however, from the few legal precedents available, that no one will own Mars. The treaty, always providing it is observed, denies the territorial claims of anyone to a celestial body. This is why the United States does not own the moon by right of exploration, as would have been the case had the treaty not been signed two years before the *Apollo 11* lunar mission. The ancient privileges of exploration do not apply to the solar system. Whether that will remain the case is somewhat doubtful, considering history.

The so-called Moon Treaty, as yet unsigned by the United States, is considerably more specific. It is concerned with mineral rights of nations who mine resources of the moon, Mars, or any other body of the solar system, along with a dozen other potentially knotty problems. One of the most often cited drawbacks to the Moon Treaty, at least for the Western world, is that there is wording in it that seems to preclude or at least hinder the development of outer space, including Mars and the moon, by private enterprise. The sections that seem to do that were introduced, it is said, by the Russians, who failed in 1962 to get all private enterprise banned from space.

Whether the Moon Treaty or some later version of the same thing is ever ratified by a majority of nations — only the United States and the Soviet Union have much chance of establishing permanent bases on either Mars or the moon in the foreseeable future — the politics and legal battling may only begin with trying to decide if Deirdre or any of the others born on Mars is a Martian.

If the colony on Mars becomes large enough, and there is every reason to think that it will, there will eventually be a movement to free Mars from the tentacles of mother Earth. Since we seem condemned to repeat the mistakes and misfortunes of history, perhaps it will come to war: the real "war of the worlds." Who knows? Maybe H. G. Wells was more prophetic than we thought. Martian spaceships may one day descend on Earth to battle the inhabitants. Only the Martians will be ourselves.

7 GIVE ME A LONGER LEVER

THE SKY IS a much fainter salmon-pink.
The dust storms do not roam; they no longer
girdle the entire planet for weeks. What winds
there are seem gentle breezes compared to those of the Mars
we knew in the twentieth century. Water flows in the deepest
parts of the fantastic canyonlands; there are small bodies of
water, lakes, in the lower regions. The summer days are colder
than on Earth but are no longer the bitter Arctic temperatures
we have so long associated with the Red Planet.

Ten people dressed in woven metallic jumpsuits stand
on a concrete pad outside a small but growing scientific station
near the Martian north pole. In the distance, a large, black
object looms in the sky, humming faintly. It is a dirigible, a
vast airship filled with hydrogen produced by electrolysis from
Martian water. Along its upper surface, hidden from view of
those on the ground, are rows of silver-black solar cells that
provide power for the motors.

The sound rises, growing louder. The airship comes in,

drifting slightly in a wind from the east. The motors growl, coming to full revolutions, the turbine-like props churning the Martian air. An airship does not land like an airplane. It never achieves negative buoyancy, except by accident. It lands by being driven toward the ground by the motors and by the elevators on the tail.

Drifting crabwise in the wind, the airship is nosed toward the concrete pad; the ground crew seizes ropes and ties it down to the mooring eyes set into the concrete. A load of supplies and scientific equipment is lowered from the gondola cargo bay onto small trucks. The crates are taken to the base headquarters.

This afternoon supply run has taken place a hundred times on Mars in the small communities of the planet. It will take place a hundred or a thousand times more in the decades to come. What is new today is that each member of the ground crew for the dirigible is wearing an improved Mars-suit. It most closely resembles the flight-jumper much favored by the air forces of the distant Earth. Instead of a full spacesuit helmet, each crew member is wearing a lightweight, spun metallic covering resembling a World War I flying helmet. On the front is a light, transparent mask not unlike a scuba diver's mask. They have lightweight oxygen bottles on their backs, but they are much more comfortable than the pioneers who explored Mars in the early part of the twenty-first century.

These men and women who are unloading supplies will cart their treasures to a base that is only partially buried in the soil of Mars. Some of their domes are in the open. So, too, are the people, from time to time; they work and play in the open more often with each passing year.

It is not quite like Earth yet, this unfinished planet; but it is becoming more so with each succeeding decade. It will one day resemble home so much that the colonists will walk around with no environment suits, no oxygen flasks on their backs. They will stand beneath that strange Martian sky and breathe deeply, filling their lungs with an air that will not kill them, not leave them gasping, curled into fetal position on the Martian sands.

If something could be done to change Mars in a fundamental way, the colonists will not be restricted forever to their rabbit warrens and modules of an underground world. They could walk outside, even if it were necessary to be partially protected from the still-harsh Martian climate. We are a race bred to stand in the open air when we want to. We may be creatures of cities in our civilized state, but they are not underground cities, and underground cities probably never will be appealing to us for any length of time.

This will be the dream of the new Martians: to live as other humans have lived in the thousands of centuries before them — beneath the bowl of the sky, listening to the haunting winds, feeling the first brush of winter frosts, breathing air deeply. They will have the dream before they first land on the planet and will pass it on to their children, and those children, in their turn, will continue it.

Transforming Mars will be a gigantic feat of fantastic scope. Nothing humans will ever have attempted will rival the engineering of the problem. Nothing we have ever done will have the complexity and at the same time the brilliant simplicity of the solution. The scientific staff on Deimos will have been at work on it through most of the colony's lifetime.

It is a profound hope, far-reaching in consequences, permanent in nature, that the men and women of Mars and their children will not have to live forever beneath the hills or forever wear spacesuits to keep out the cold and noxious air. The alternative, if it were possible, has very limited appeal: to change humans to fit the conditions of the Red Planet.

With advanced genetic tinkering one of the four types of nucleotides in the DNA could be substantially changed, resulting in human beings genetically engineered to accept the Martian climate and air as wonderful, balmy, and invigorating. Maybe fantastic things could be done with the DNA's polymerase, the protein responsible for making new DNA. Perhaps some synthesis of glycopeptide, the natural substance that protects fish from freezing in 29°F water, might someday be adapted to humans; we might then find a − 140°F day on Mars quite comfortable. Possibly some form of adaptation through genetic tinkering might be found that would allow human lungs to extract the tiny amount of oxygen in the Martian air — who knows? Perhaps mankind could be turned into CO_2 breathers to whom the carbon dioxide of the planet would be as refreshing as a Colorado mountain breeze.

With such major changes we would indeed become the Martians because we would hardly resemble humans. But this sort of transformation is an unlikely event. We humans have become used to this frail shell we call a body; it may not be the best design around, but it has served us well and it is doubtful we will be very quick to discard it for another, even if the new one would allow us to live unprotected on the face of a planet as harsh as Mars is now. It is much more probable that we will seek ways to transform Mars into a world more to our liking. Doing this sort of thing was once called "planetary engineering." It has also been called "planetary ecosynthesis," but mostly it goes by the name "terraforming."

Creating new worlds where only nature's failures existed before is the human race taking on the attributes of gods. But all of our modern technology, our whole modern world, would be a fantasy of the gods to a man born a few centuries ago. Messing around on a gigantic scale with an entire planet is something new in magnitude. How a planet could be modified so that we might live there comfortably someday taxes even our twentieth-century imaginations.

Why is Mars a marginal planet for the human race? Why is Venus too hot for us? Of the nine planets and forty fair-sized objects in the solar system, only Earth has flourished and supported life. Was this entirely accidental, the result of natural consequences, or was it something more? The answers to these kinds of questions are important, because without them we cannot look for ways to modify a planet to make it more livable.

Planetary systems are believed to form as a natural by-product of the birth of stars. This theory was first advanced by Pierre Simon Laplace in the eighteenth century. A star is created when a cloud of dust and gas collapses under its own gravitational pull. The embryo star forms in the center of the cloud and is surrounded by a vast, rotating disk of leftover matter from the formation. It is from this leftover matter — gas and dust — that planets are born.

We now have what we think are samples of this original dust. Microscopic amounts of it were trapped in meteorites that were formed out of the stellar dust 4.6 billion years ago. Some of these dust history books have amazed us all: exotic isotopes have been found in the dust that could have been created in a gigantic supernova explosion, the death of one star attending the birth of another. The dust that formed our sun and our solar system once contained — and still does, by direct lineage — star dust from ten to fifteen long-vanished solar systems.

A miniature version of the process is visible in our solar system, in the rings of Saturn. Studies in laboratories on Earth have shown that the mechanism that formed the rings of Saturn is very much like the process that must have gone on in some primordial past of the solar system. We have been able to study the rings of Saturn and their odd construction, gaps, and intricate rearrangements to such a degree that many scientists feel similar events must have formed our solar system, proving Laplace's hypothesis of nearly two centuries ago.

The dust grains collide and eventually attach themselves to each other, forming a great carpet of solid particles in the same place as the proto-star disk. They grow larger after each tiny collision until they too have enough internal gravity to attract more material quickly. Like a cosmic vacuum cleaner, the tiny proto-planet sweeps space for grains and more dust. The mass eventually collapses into a sphere of relatively dense, hard material much smaller and much colder than the newly formed star; the planetary body never becomes large enough to cause the process of fusion that would make it into a sun.

The volatile gases on the nearer proto-planets are evaporated by the heat from the young star, and what remains are solid, rocky worlds of heavy matter. The planets farther away from the star collect large amounts of gas from the surrounding thin cloud that remains and their distance from the star keeps the volatiles intact in their gravity fields. Much farther out, the planets are so cold they freeze over shortly after formation; and on the very edge of the embryo solar system, the gases freeze into ice, producing a cloud of cometary debris.

This natural process of star and planetary birth seems to be fairly common throughout the galaxy. Computer simulations have shown that most single, slow-rotating stars — those like the sun — may have planets. A solar system will typically be two, three, four, or five solid, dense bodies and perhaps an equal number of giant gas bodies with a few remaining worlds completely frozen over. In our own solar system, we call these objects Mercury, Venus, Earth, Mars, Jupiter, Saturn, Uranus, Neptune, and Pluto — our planets.

Our solar system too has some leftover debris: a selection of moons for the planets, asteroids, comets, meteors, and assorted flotsam and jetsam.

What happened to the bodies after formation also seems to have been a by-product of natural processes: the solar wind pushed all the remaining gases from the system, and the residual dust grains and chunks of gravitationally associated materials rained down on the unprotected surfaces of the rocky planets in a cosmic barrage. This produced the cratered terrain that has been seen on all the planets in the inner solar system: Mercury, Venus, Earth, Mars, the moon, and most likely the asteroids as well. Asteroids have not yet been photographed by spacecraft cameras — they will not be photographed until the 1990s — but most scientists expect to see heavily cratered surfaces on them. This cosmic rain of debris is also why the tiny moons of Mars, which are probably captured asteroids anyway, are fairly heavily cratered.

Earth has been geologically active; it has renewed its surface through vulcanism and plate tectonics. The great land areas of our planet do not appear strewn with craters. But little Mercury is a corpse of a world. The millions upon millions of impacts on its surface date back three billion years or more. Our moon is also a shattered, sterile world. Mars is in between: some of its primordial bombardment is still visible; some of it has been erased.

But it is not geology alone that makes a planet have life or even be habitable. Information now available through computer simulation shows that the position of a planet in its solar system is extremely important. The basics for life — atmosphere, water, day/night cycles, and a hundred other factors — exist only in a very narrow band mainly determined by the distance of the planet from its sun.

Spacecraft have provided us with very substantial amounts

of information about the present conditions within the solar system. It is possible with computers to trace backward in time to the point of formation and get a clear picture of how the planets evolved. With Earth as a comparison point, we have been able to see why our own world went on a different evolutionary track from the others. In other words, we now know why the other worlds have no life on them. There is no reason to suppose that our solar system's situation is unique. It may have been repeated over the galaxy a hundred thousand times in the past six million years.

The two nearby planets, Venus and Mars, suffered from natural events from which Earth escaped. The first is a planet too hot, the other too cold. Only Earth is the cosmic baby bear's porridge. Had Earth been only 5 percent closer to the sun, it would have suffered the same fate as Venus did four billion years ago. Were it only slightly farther away, our world would have ice over most of its surface.

Mars also suffered from another natural process. Most of the atmosphere introduced to the planet by the volcanoes (Earth and Venus retained theirs) was lost to space. Mars was too small to hold it. Planets lose their atmosphere because the molecules move quickly and will fly off into space if the gravitational pull of the planet is insufficiently great. The lighter gases are lost first: hydrogen, oxygen, nitrogen.

Terraforming is a technique or group of futuristic engineering procedures that could be used to change a planet from its current conditions to ones more closely approximating those of Earth. Ideas for terraforming have ranged from a seventeenth-century suggestion that the craters on the moon were artificial, built by the inhabitants to provide shade during the two-week-long lunar "day," to a twentieth-century idea that someday planets could be created simply as a work of art — by a great sculptor in action against the greatest piece of marble ever. Arthur C. Clarke's haunting worlds in *Against the Fall of Night* were tiny planets shimmering in the glow of the sun, planets that a future race had created from the flotsam of the solar system as a great artistic achievement.

But until recently, terraforming has been confined mainly to the pages of science fiction. And it was a subject without much substance until unmanned spacecraft investigated the solar system and reported back the present situation on each planet. Prior to that, planetary conditions were too imperfectly known, and discussing how a planet might be changed for the better was remarkably fruitless. The key was an understanding of planetary evolution — a planet's birth and the details of its childhood. Without that understanding, it was not possible to predict the outcome of any changes that might be introduced.

A planet moved closer to the sun would be warmed in a natural way. The regolith would give up gases that would form an atmosphere. The blanket of air would trap sunlight in the same way the glass in a greenhouse does. The warming trend caused by the greenhouse effect would be planet-wide, and additional gases would be released from the soils. The effect would be self-generating and sensibly permanent.

Our own planet will be a warmer place in the twenty-first century because of an increasing greenhouse effect from burning fossil fuels, principally coal. By the year 2065 atmospheric carbon dioxide levels will have doubled, causing an expected 3–7°F rise in temperatures over our world. Among the problems anticipated for Earth are rising oceans and shrinking coastlines, caused by melting of the polar ice mantles.

The composition of a planet's atmosphere is as important for us as the density and extent. Human beings are not disposed to breathing carbon dioxide, which is the major constituent of all early atmospheres; it is still the major part of the air on Venus and Mars. The key to terraforming the air on any planet lies in what makes up the soil and the polar caps. Bodies of water — or any water bound up in the rocks and soil — are crucial. In the case of a planet with an icy crust and water-ice polar caps, heat would cause the water to enter the atmosphere; that would provide both water vapor and oxygen, since one of the major parts of water is oxygen. Soils that are basically oxides also contain oxygen. These soils will give the oxygen up to the surrounding air if they are warmed enough.

If future engineering techniques can cause a gross evolution in a planet's atmosphere and average temperature, a second component of planetary engineering can be introduced: the biological element. There is no reason or need, obviously, to wait a few billion years for some mixture of ammonia, water, methane, hydrogen, cyanide, carbon dioxide, and phosphates to be stirred in Charles Darwin's "warm little pond." We can be selective and take what life we find suitable from Earth and put it on the new planet.

There are numerous biological folk on Earth which would make efficient terraforming agents. In fact, the atmosphere of Earth evolved because of little blue-green algae; they converted the early atmosphere to the one we know now by taking in carbon dioxide and releasing oxygen. Studies by NASA over the years have indicated that the simple introduction of satisfactory biological agents to an atmosphere that can partially support Earth life now could result in breathable air in a few thousand years. With genetic tinkering, a brand of biological agent could be created that would put oxygen into an *already terraformed* atmosphere — one dense enough for life, but with air made up of the wrong gases — in a fairly short interval.

If all this sounds slightly impossible, imagine a seven-teenth-century man standing amid the forest of Manhattan Island and having someone tell him that the trees would be gone in a few hundred years and a building more than a thousand feet high, with people working in it, would occupy the site.

Terraforming ideas are like a drug: once they take hold, there's no escaping them. The word "terraforming" was coined by Jack Williamson in two short stories published in 1942, but the basic idea goes back much farther, to an Olaf Stapledon book in which Venus was denuded of her oceans (which don't exist anyway, but the author didn't know that in 1930) by generating oxygen through electrolysis.

The principles involved are sound: put a planet on a different evolutionary track than the one it is on now and it will become a different planet. As terraforming is now conceived, it is a group of related technologies and proposals by which mankind can think about expanding into the solar system in a wave of migration that may one day reach the stars. It is a far grander concept, larger and more fantastic, than building cavelike colonies in orbits between Earth and the moon, the "space colonies."

But it is also a brutal concept. It is the wholesale changing, presumably permanently, of an entire planet by restructuring the way its climate operates and modifying its other natural characteristics.

An Earth-like atmosphere on Mars, for example, from a strictly Martian standpoint, would be classified as pollution. A hardcore terraformer would not think twice, however, about creating one by almost any means possible. Not surprisingly, by some people's definition terraformers have horns, always wear black, and mix up concoctions the human race might be better off without.

Looking at Mars, then, with the gimlet eyes of a terra-former, we can see a wide range of possibilities open up before us. But before taking any steps, even on paper, it is worthwhile to run a balance sheet of what is good and what is bad about the present Mars. That should tell us what we — and our future colonists — would like to change.

On the good side, Mars has nearly as much land area as Earth. This fact seems on first glance to offend common sense, because Mars is only a little more than half the size of our planet. But it has no oceans, so the land areas of the two planets are approximately the same. Mars has a day almost the same length as ours, and the tilt of the axis is presently about the same, too: 25 degrees. This means that, like Earth, Mars has seasons, but because Mars is farther from the sun and moves more slowly in its orbit, the seasons are twice as long.

The orbital tilt has not, however, always been the same. It evidently varies from 14 to 35 degrees, with the variations coming at intervals measured in tens of thousands of years. The variation is caused by the lack of a large moon; Earth's moon provides a gravitational balancing effect for our planet's axial tilt. For now, the axial tilt of Mars is stable.

As on Earth, there are mountains, flat plains, sand dunes, clouds, and wind. If a stretch of Martian plain were covered with cactus, creosote bush, and mesquite, we couldn't distinguish it from a desert in northern Mexico (except, of course, for the pastel-pink sky).

The bad side we already know: Mars is too cold, there is little atmospheric pressure (equal to the pressure at an altitude of 20 miles above Earth), and the atmosphere has the wrong composition. In addition, there is not much water vapor in the air and little, if any, "free" water on the surface — no lake or even a pond. On most of the surface — but not all —

the atmospheric pressure is not sufficient for water to exist in a liquid state.

The first question for terraformers is, What can be done to change the Martian conditions to those more approaching Earth's? The second question is, How can we accomplish it? When the questions are considered together, it turns out that there are a number of ideas that can be discarded for one reason or another — some because they are basically impractical or uneconomical, or because they would damage Mars too much for too long and couldn't be done while people were living there; others because they are beyond the technological capabilities of the first century or two of the colony.

That eliminates a few of the favorites of science fiction writers. One discard is setting off a fusion reaction on Phobos to create a mini-sun to warm the planet. We have no real grasp of how to do such a thing, since fusion has yet to be achieved in the laboratory. Even if there is a fusion breakthrough in this century, we will still be a long way from being able to put a shiny box of some kind on the inner moon and "poof," here comes the sun.

Also eliminated is an idea from a 1975 NASA report on the habitability of Mars in which it was suggested that an atmosphere could be imported from somewhere else, presumably from the outer planets, Jupiter or Saturn. It could hardly be cost-effective, even if there were some way to do it. Terraforming is such a mind-stretching concept that it can get out of hand easily.

Another bad one is the notion of raining icebergs taken from the rings of Saturn onto the Martian surface. While it is true that one of the objects 40 miles or so in diameter would dig a hole about 25 miles deep and that the atmospheric pressure at the bottom of the hole would be almost 500 millibars, it's not a terribly practical plan. Still, the 500-millibar pressure is about twice what human beings would need to feel comfortable.

Advocates of the Saturn iceberg plan, and there have been more than one might think, also point out that the icebergs would disintegrate on impact, throwing water vapor and gases into the atmosphere, thereby increasing the net effect. Not only that, they say, but the impact would throw up native volatiles too, including water vapor from the exploded Martian permafrost. The purpose of changing Mars, however, is to get the colonists *out* of a hole, not to put them into a deeper one. Even if people would be comfortable in the iceberg pit as far as atmospheric pressure goes, they would be farther from sunlight than ever.

We can eliminate another fictional thrust at the problem made in 1976: to trigger the Martian volcanoes with atomic bombs. The idea was that the volcanoes would begin "outgassing" (a planetary burp; to continue the analogy, baby planets do it a lot), and the resulting concentration of gases in the atmosphere would make it thicker. This would help trap the solar heat, which would lead to a planet-wide warming trend, which would lead in turn to a melting of the permafrost. The extra water vapor would cause the atmosphere to be thicker still; the extra heat that would be trapped would increase the rate at which CO_2 left the ice cap and went into the atmosphere. Eventually the process would be self-sustaining and Mars would be a much nicer place to bring kids up.

The drawback of the plan, in addition to violating what would probably be innumerable nuclear treaties past, present, and future, is that the Martian volcanoes may have been dead for billions of years. The chance that any number of atomic bombs would do anything but move the landscape around and cause radiation hot spots on the surface is close to nil.

Despite the drawbacks in the application, there is nothing wrong with the intended *result*. Mars does need a thicker atmosphere to start a process that will warm up the planet. The process, the greenhouse effect, will occur on Mars because certain gases, carbon dioxide and water vapor among them, absorb heat from the surface of a planet and send it back down again to warm the ground. The greenhouse effect from CO_2 is relatively small; that from water vapor is quite large.

If we run a computer simulation of what would happen to Mars if the water vapor were increased, we find at first what seems to be bad news: with only one-third the surface gravity of Earth, to get the same atmospheric pressure we have, the atmosphere has to be much denser and the greenhouse effect must be much greater than here.

The good news is that this is a "balance out" proposition in which a one-bar atmosphere (Earth's is about 1,000 millibars, or one bar) is just about perfect for Mars. If a one-bar atmosphere were in place, the planet would be much warmer than it is now and would stay warm with the ordinary amount of sunlight that reaches it. The greenhouse effect would be permanent.

The question for terraformers is how to send enough water vapor into the atmosphere to start Mars on the way. If extra energy — in the form of heat — amounting to *one percent* of the sunlight now reaching Mars could somehow be generated, in three hundred years the planet would have a substantial atmosphere. If by some fantastic feat of engineering an amount of heat equal to the sunlight hitting Mars could be produced and directed onto the water sources, the effect would happen in only three years. It is important, however, to maintain perspective on terraforming. The three-year plan would involve vaporizing water equal to a depth of ten feet over the entire surface of Mars.

Heat is obviously the answer, but how to generate it and

rain it down on the surface of Mars is a healthy engineering feat. It just wouldn't do to think up ways of temporarily increasing the output of the sun. That would work for Mars (no one, however, has even the foggiest notion of how it could be done), but it would be disastrous for Earth. Any increase in the sun's production of energy big enough to *really* benefit Mars would fry our own planet.

Instead of trying to increase the heat coming to Mars, we could try to use what is available. This might not be as difficult as it sounds. The poles of Mars, which are mainly water-ice (when heated enough, water-ice would become water vapor; this is what we want), reflect 77 percent of the sunlight reaching them. Like the ice caps of Earth, they are bright white. If they were darker, they would absorb more sunlight and the temperature might be high enough for the ice to start melting, releasing water vapor into the atmosphere. If the reflectivity of the caps could be reduced from 77 percent to 73 percent, this seemingly small change by itself would trigger a greenhouse effect and a much denser atmosphere on Mars in about a century.

After separating what we might be able to do from what is barely possible, the answer to terraforming Mars is that it can be done. We will have to be dedicated. We will have to wield a technological hammer of a high order against the planetary anvil; but if we want to badly enough, we will one day step onto the Red Planet and have something a bit more like home.

8 IT'S A ROTTEN JOB, BUT SOMEBODY'S GOT TO DO IT

THE COLONY SHIP is a great trefoil, a gigantic **Y**. It makes the original first expedition ship to Mars look like a Wright Brothers airplane compared to a 747. Most of the colony ship's components will be boosted into Earth orbit by VHLVs, Very Heavy Lift Vehicles — the same kind developed for moving large weights into orbit during the space station era. Once all of the components are in orbit near the space station, each arm of the **Y** will be assembled, checked out, and prepared for flight.

The arms of the trefoil are interplanetary spacecraft: each has living quarters developed from space station technology, life-support systems also developed from the same source, communications equipment, and power modules designed to use sunlight converted to electricity. A Mars lander is attached to the arm. It can carry five crew members down to the planet (or bring ten up in an emergency). The landers can carry about ten tons of cargo.

Each of the independent spacecraft-arms will be attached to conventionally powered (liquid oxygen, liquid hydrogen) booster stages that use engines derived from the current space shuttle. When the colony ship departs from the orbit near the space station, the giant booster stages fire for about fifteen minutes. The boosters then separate and each of the spacecraft continues toward Mars. After two days, they rendezvous and dock, and during the next few days, they are assembled into the final configuration for traveling to Mars — the trefoil.

The colony ship begins to spin on the sixth day and the people aboard will feel artificial gravity. If the great ship rotates only one-third to one-half a revolution per minute, the artificial gravity in the living quarters will be 0.3 of Earth's — quite comfortable for the crew and not too much less than the gravity of their destination so they will have plenty of time to adjust to a gravity substantially less than the gravity of Earth.

Each colony ship can carry fifteen new Martians to the permanent settlement. When they have arrived on the surface the Mars landers will be fueled with carbon monoxide and oxygen generated from the Martian air. Every two years, each colony ship — the trefoils — will bring fifteen more people to the settlements of the Red Planet.

As the *Lowell-1* spins on its journey from the blue and white planet to the red one, one side of the pinwheel faces the sun. On this side of the great craft are the communication antennas, solar-electric generators for power, fuel tanks, engines, and maneuvering propulsion systems. On the opposite side, the shadowed part of the pinwheel, are the crew's quarters, cargo holds, and landing craft. Always in the shadow of the sun, with the bulk of the spaceship between them and any potential solar flares (a sufficiently strong solar flare could fry an exposed crew with lethal radiation), the crew waits for the moment they will call traffic control on Deimos for docking instructions.

As they go in from the black of space to the red glare from Mars, they can see in the distance the giant solar mirrors that are pouring sunlight onto the Martian surface. Those with keen eyes or access to one of the few optical telescopes aboard the colony ship may be able to see the domes of the "shadow makers," as the scientists who man the mass drivers funneling dark material from Deimos to Mars are called.

Mars is being changed, even as the colonists arrive. It has been so from the very beginning, although the process is a much greater and grander effort now. The keen-eyed of the colonists would be able to see the darker portions of the polar regions where the material from Deimos has landed. They might be able to see the wisps of water vapor coming from the poles as the solar mirrors heat the surface.

The first approximation at terraforming Mars will be with mirrors, sent from Earth. There is nothing in the technology involved in building mirrors and structural supports that is beyond our current abilities, knowledge, or perhaps even funding. Using giant space mirrors is not even a recent concept: the rocket pioneer Hermann Oberth wrote about them in the 1920s. He in turn was evidently inspired by the ancient legend of the Greeks in which light was reflected from soldiers' shields to set fire to the sails of enemy ships in the harbor.

Mirrors for use in space will be flash-coated with a reflecting metal. The process is relatively simple: a solar furnace would provide heat and the heat would be delivered to an evaporation gun. The gun would be used to focus a conical molecular beam on a mold, perhaps a giant Mylar film; the mold would be sprayed by the molecular gun and gradually a thin metal film would lie on the Mylar surface. It would be a giant mirror.

The designers of the structural support needed for the thin mirror could take a cue from some of the racecar constructors: rigid beams made of sodium-covered graphite epoxy and carbon fiber. The actual mirror part would be sodium-covered Mylar or Kapton, another thin, sheetlike alternative with perhaps more resistance to damage.

The Mylar or Kapton would come from Earth, shipped to a construction orbit by the space shuttle. The sodium would be flashed onto the mirror after it was put together in space by a construction crew operating from a space station. We could build a mirror — maybe two — even before the space station is built; astronauts working out of the space shuttle's cargo bay could put them together by using the manned maneuvering unit already prepared for shuttle exterior work.

Constructing the mirror in Earth orbit might be interesting, though, since the pressure of photons would tend to make it depart outward into the solar system. It would be a solar sail hardly different from those that would ship cargo to Deimos. A small chemical-rocket package would be necessary to keep the mirror in position because the photon stream from the sun is irresistible and constant.

Upon completion, the great mirror would be "released" into space (meaning the orbit-adjusting rockets would be shut down). It would take off like a helium balloon released from a child's hand and journey to Mars by the pressure of light with no fuel costs. This *Genesis-1* mission could conceivably leave Earth orbit as early as the late 1980s. *Genesis-1* would be a robot mission; it would establish an orbit in exactly the same way our spacecraft have done since 1971. The robot craft would computer-direct the mirror over Mars to focus sunlight on a wide area of the polar regions. Exactly where doesn't even matter. The solar heat focused by the mirrors would start melting the polar caps and thawing out the permafrost regions long before the first expedition lands on the planet.

By the time of the first landing, the mirrors would have been heating the polar regions for almost ten years. If the mirrors turn out to be cheaper to produce and send than we currently think, then four or eight or twenty could be in orbit by the time the first permanent base is established on the planet. By the time Deirdre is living in the first city on Mars, there would be some modest changes going on.

At first the amount of water vapor entering the atmosphere will be tiny; but as more and more terraforming techniques are applied, the boil-off will increase in force and volume, until the atmospheric density rises from a good vacuum to a pressure where water can remain a liquid under the limp Martian sun. As the polar caps give off their water over the next fifty years, the extra water vapor will start a greenhouse effect that will leave Mars with an atmosphere that has some dignity to it, if not acceptability to humans. The planet will warm, very slowly at first, but as the greenhouse effect grows, so will the warming.

When Mars is warm enough, the large regions of permafrost will melt and more liquid water will leave the planet for the air. On the warmer world, the winds flowing from the equator to the poles will increase in size and strength and carry with them the extra solar radiation in the form of heat, which the improved atmosphere will now retain. Water vapor will be pouring into the atmosphere from the great Tharsis glaciers in the volcano region of Mars. The dust storms will no longer sweep the planet in quite the way they did for tens of thousands of years; they will no longer carry ten times as much dust into the air as on the worst dust-smog day in Albuquerque. They will now occasionally carry moisture, and the myriad tiny dust particles in the Martian air will fall occasionally as mud clumps surrounding a genuine Martian raindrop.

Regrettably for the Martian colony, the single fact of the soil's ability to be moistened and give off trace oxygen will not suddenly transform their planet; the whole surface of Mars, even if it were rained on for weeks, would provide only a few percent of the oxygen required for an Earth-like atmosphere. The days on the equator at high noon in the middle of the hot season may be a balmy 75°F. The nights will still be very cold, a hundred below or more; but the drop at sunset will never be as great as it once was, because the thicker air will retain more of the heat of the day.

Deimos, in addition to functioning as a supply and scientific base, will become a center of terraforming activity. A mass driver — a modern name for the old "electromagnetic space gun" much beloved of science fiction writers — will be showering Mars's polar and permafrost areas with the moon's dark dust and grit. The device will look a little like a conveyor belt made of buckets. Magnetic impulses driven by solar-generated electric energy will accelerate the small buckets holding the mass driver's payload of dirt — or anything else.

By the time Deirdre takes her subway ride to the laboratory where she works, Mars will have changed from the days of Viking. The air will still be a poisonous mixture that humans can't breathe (mostly carbon dioxide) but to the people of the first city on Mars, it will be their atmosphere, or simply *the air*.

One of the projects Deirdre will work on in her laboratory is the plan for seeding Mars. In her time, some of the planet will have been seeded with tiny creatures that only require trace oxygen to do their simple-minded task of photosynthesis. The dirigibles used by the colony will have begun the seeding as soon as the temperature and air pressure are sufficient in the deepest, low-lying regions to allow the survival of biological terraforming agents. The smallest and most lowly of God's creatures killed the Martian menace in *War of the Worlds*. Very different, but equally tiny and lowly, will be those creatures that make a home for the new people who call themselves Martians.

"It is not enough that a thing be possible for it to be believed," wrote Voltaire; and he was right. Terraforming is a wonderful, staggering concept, full of hope and optimism and an almost Victorian belief in the potential strength, the benefits, and most of all the rapid progress of technology. But is it truly something a future generation can do to a planet? More specifically, can it really be done to Mars? Can we make it rain once again on the Martian deserts; will we one day walk in the open beneath the strange pink sky without suits of any kind, with no protection from an environment that would kill us now in a few seconds?

The two best gases for creating a greenhouse effect are water vapor and carbon dioxide. Fortunately for terraformers, Mars has plenty of both. The polar caps are water-ice and the atmosphere is 95 percent carbon dioxide already, so the ingredients for terraforming are there.

Estimates vary considerably, but most put the thickness of the residual northern ice cap at half a mile. The residual cap is that part of the polar ice sheath that never goes away, even during the Martian summer. It contains somewhere between a thousand and perhaps fifty thousand times as much water as exists now in the atmosphere in the form of water vapor. The southern polar cap, which was formerly thought to be mainly carbon dioxide, is now believed to be water-ice too, so the total amount of locked-in water on Mars is very substantial. It has never been a "dry" planet, in the larger sense.

The present nature of the caps is handy for planetary engineering projects in the sense that both caps alone contain enough locked-up gases to change the air in a major way, if the means can be found to release them. There is also substantial water locked into permafrost regions, frozen in place by time and climate; the largest permafrost region seems to be located at all latitudes poleward from 46°N and 35°S. This frozen water layer is buried one-half inch to four inches deep in the soil.

If enough water vapor could be added to the atmosphere to increase the total pressure by only 10 percent, the increased temperature over the entire surface of Mars might rise as much as 18°F. The same amount of carbon dioxide added to the air of Mars would change the temperature only about a hundredth as much. Water vapor is much better than CO_2 for greenhouse effects because it absorbs solar radiation in better — and more — parts of the spectrum.

Thus two to twenty mirrors in orbit, reflecting solar energy onto the poles, could start the changes on the planet so that humans could live there someday without space-suits — without even Mars-suits. The mirrors would not have a major impact on the Martian air in hundreds of years if nothing else were done at the same time. But other things *can* be done to hasten the process, including dropping dark-colored dirt on the poles and permafrost regions to absorb more sunlight.

The effect of throwing dirt at the planet is the same as that of wearing a dark coat in summer: things get hotter when a surface is darker. The surface of Deimos, as well as of Phobos, the other moon, is composed mainly of a deep layer of dark dust and pulverized rocks known as carbonaceous chondrite. "Carbonaceous" is the key word, because carbon is black. The black layer from Deimos would eventually cover the polar plateaus and the permafrost regions of Mars, which would then absorb many times more sunlight than the surface does now. The result will be about the same as putting an electric blanket on a block of ice.

A mass driver on the other moon could vastly increase the rain of dark debris, and it would be relatively simple for the crew of the Deimos station to reach Phobos with the necessary equipment.

In time, the combination of the mirrors and the dark rain would create an atmosphere with three times the mass of Earth's atmospheric mantle, and a one-bar pressure, which is quite substantial by Earth's standards. The currently accepted figure for the atmospheric pressure on Mars as it is today is 7.5 millibars.

But even twenty mirrors and two mass drivers chucking the sun's rays and dirt on the Martian surface are not going to get the job done in a time frame the colonists will find acceptable. One way to speed the process up is to build more mirrors out of native materials and put them in Mars orbit. They could be assembled from materials sent to Deimos from the Martian surface. Mars has abundant supplies of sodium, a useful ingredient in mirror construction. It also has plenty of magnesium, which would also be entirely satisfactory in making mirrors.

The mirror method of terraforming Mars was initially met with considerable pessimism because the number of mirrors required for a substantial change is huge. The cost of building so many on Earth or in near-Earth space and then shipping them to Mars seemed prohibitively high. As soon as it was known that sodium and magnesium were abundant in the Martian dirt, part of the problem was solved. But getting any mirrors or their structural parts of native manufacture up into orbit from the Martian surface remained a looming problem.

No Martian colony is going to be making spaceships anytime soon. Whatever ferry-type vehicles are used on Mars will have to be built, bought, and shipped from Earth. Leaving aside the technology required, the cost of constructing any sort of heavy-lift vehicle designed to get Mars-made mirrors into orbit has always been too high to be considered seriously.

There is, however, the space elevator, the stepchild of an older concept called a skyhook, which was supposed to be a magical device for avoiding expensive space freighters, the only way known for years of transferring cargo from the surface of any planet to an orbit above it. In its simplest form — it is inevitably one of the simplest "magical" engineering feats — it is a cargo line dropped all the way to the surface from a satellite in orbit.

The orbit would be geosynchronous with the planet below, meaning that the satellite would hang over one spot indefinitely, using on-board propulsion corrections from time to time. This is the same sort of orbit that is used today for telephone and television satellites over Earth. The on-board propulsion system would also be used to return the anchor satellite to its original position after a cargo had been lifted (the anchor satellite would lose altitude during a lift).

An electrically powered cable car would run up and down the line carrying cargo; the electricity to power the arrangement would come from a solar-powered satellite. If safety factors could be brought up to the point where anyone thinks it is safe to ride one, a space elevator could carry passengers as well as cargo.

There are several hitches to the space elevator, at least

on paper. (No one has ever tried to make one.) First, the material of the cargo line would need to be many times stronger than anything available on Earth today. The key words, however, are "today" and "Earth." New structural materials are being created all the time. It is expected that new alloys and very strong materials may come from the experiments in space manufacturing aboard the spacelab and the space station over the next ten or fifteen years. It is entirely possible that a cargo cable of the required strength may be available to the colonists on Mars.

A more bizarre form of the space elevator is the "kiss and run" cargo carrier. This is an anchor satellite around which a cargo cable would be wound. The length of the cable and the altitude of the satellite are equal, so the cable would come down to the ground once each time the satellite revolved around the planet. A watcher on the planet's surface would see the skyhook come down vertically for an instant and then suddenly disappear into the sky once more. For the brief moment the cable touched the ground, an empty cargo module would be dropped and a full one snatched. If the revolution period of the anchor satellite were doubled, a cargo line of the same length would touch down on the surface twice a day, on opposite sides of the planet.

If the space elevators could be used on Mars, they would replace any other kind of freight-carrying vehicle. It would be simple to move supplies and people down to the surface and mirror parts up.

No matter how the terraforming of the planet is accomplished, it will be the most stupendous engineering project of any century in human history. Each century has its engineering challenge, its "wonder of the world," and each challenge has usually been much more difficult than the ones before. A seventeenth-century engineer not long emerged from the middle ages would have been baffled at the prospect of building a steam tractor. He would have measured his Total Energy Budget for the project in horsepower, all right, the real kind. Nicholas Cugnot invented the steam tractor a century later.

An engineer in Cugnot's time would have thrown up his hands at the complexity of building a canal through the narrow, 100-mile-long isthmus separating the Mediterranean from the Red Sea. But in 1869, Ferdinand de Lesseps did just that, a century after Cugnot. He made considerable use of steam power, and so did the ships that sailed through after it was built.

Those wonderful engineers of the late nineteenth century who built the Suez and began the Panama Canal, and the engineers who followed them into the turn of the century, the ones who built the great trains that ran in Europe at speeds faster than Amtrak does in the United States today, would have been equally baffled by the prospect of putting a man on the moon. They had conquered distance in three-quarters of the world, had ships that could visit any part of the globe; they had the telephone, the telegraph, the earliest radio, submarines, and enough armaments to make very respectable wars. They did not have a liquid-fuel rocket. Yet a group of engineers landed us on the moon exactly a century (to the season of the year) after the two gigantic ditches of the Suez Canal were finally connected.

In our time, we think the achievement of having a space freighter that can carry 65,000 pounds into an Earth orbit is incredible. We have giant communications antennas 22,500 miles out in space. We have sent spacecraft to almost all of the planets known to the ancient astronomers, or even to the astronomers of the nineteenth century. And even our engineers are baffled by the complexities of terraforming. We know it could be done, but we don't quite have the gear-up technologies in place, yet. And we hear the same old cry that has echoed down over the centuries: "How will you get the money to do that?" The usual forms of planetary engineering — mirrors, dark material on the poles, seeding by oxygen-producing bacteria — must eventually have some economic base from which the new Martians might finance the changing of their planet. One of the ways it might be done — in fact it might pay for the whole job — is through the export of raw materials to Earth.

We live on a world with finite resources. We're rapidly running out of some minerals, and others are in short supply. Aluminum, for example, a substance of which we use a tremendous amount, is presently produced from the mining of bauxite. Bauxite has about 25 percent aluminum by weight but supplies of it are slowly decreasing. Companies such as Alcoa have recently been mining the ore anorthite, which has 19 percent aluminum by weight, but it too is not inexhaustibly abundant. Looking elsewhere, however, now that the space age has advanced as far as it has, we find that the moon has a very substantial amount of anorthite. Titanium, another light metal we use in quantity, is extracted from an ore called rutile. It can also be taken from ilmenite, however; and it turns out that the moon also has plenty of that, mainly among the basalts of the lunar seas.

This discovery, that the moon was not unlike Earth in terms of mineral resources, was a major event. It confirmed that the moon was not made of green cheese; it was made of more or less the same stuff as Earth, sometimes in different proportions. Because spacecraft to the moon and elsewhere, including Mars, have so filled our dance card with information about the early years of the solar system, we now realize that the raw materials the Earth could use are all around us in space. It isn't empty space, the great void we thought it was.

The situation is not too different from that of the oceans: it's an inimical environment to work in, but it can provide an enormous wealth of resources. All we could ever need.

In the first rush of fascination over raw materials from outer space, the moon was the logical candidate. After all, it takes only 5 percent as much energy — propulsion — to send a payload from the moon to an orbit around Earth as it does to ship the same payload from Earth's surface. Obviously, many people reasoned, as civilization spreads out into Earth's higher orbits in the long-predicted "space industrialization," raw materials from the moon will be cheaper for construction.

Someone has to mine those minerals and someone has to send them to the construction orbits. So was born the fantasy of "moon miners of the void," in which a permanent base on the moon would use mass drivers to launch mineral payloads to waiting "mass catchers" at building sites. The moon's resources would "finance" the building of lunar bases and especially space colonies in one of the Lagrangian (such as "L-5") orbits near the moon. Three large mass drivers over ten years could ship enough stuff to build a space colony in which more than ten thousand people could live.

There is nothing wrong with the lunar mining scenario, and it may come about in the twenty-first century. The resources would be useful for a space civilization, which certainly couldn't be a petroleum-based society, as ours is now. Civilizations in space would use solar and nuclear energy for power. They're more plentiful and much cheaper to get, since oil is most unlikely to have formed on the moon or Mars or anywhere else but Earth. Oil requires some dead dinosaurs to start with, more importantly what they ate; and there was none of that on any of the other planets or their moons.

While lunar ore is still a very viable candidate for being mined, it may not be the best quality or the best quantity and it may not be the cheapest, either. There are the asteroids too. There may be more mineral wealth in the asteroids than on the moon.

The belt of flotsam and jetsam of the solar system lies in a great ring, 340 million miles in diameter, between Mars and Jupiter. The little planets, as they are sometimes called, are indeed small. The largest known is Ceres, with a diameter of about 600 miles. The smallest asteroid would be defined as a meteor.

Asteroids and meteors are the leftover building blocks of the solar system. The original belt formed along with the rest of the planets as a cloud. There are fifteen to thirty major bodies from 100 to 600 miles in diameter. The bodies played cosmic bumper cars for millions of years, leaving the rubble we see. A few of the larger pieces remained intact, such as Vesta, which is spherical and about 312 miles in diameter — truly a little planet. Few of the others are spherical. Most are irregularly shaped chunks; the average body looks much like Phobos or Deimos. Most of them rotate in between thirteen and six hours so the "day" on one of the little worlds is that long.

Since asteroids come from the original compounds that formed the larger planets, it can be presumed that somewhere in the belt are worlds made up of just about anything in the periodic table of elements, including the metals nickel, iron, platinum, gold, chromium, iridium, osmium, palladium, and rhodium. There are some asteroids that astronomers believe are water-ice, an obvious source of oxygen, hydrogen, or water for civilizations in space.

Asteroids appear only as points of light in the most powerful telescopes, so our information about them is indirect; our ideas about their composition come mainly from our newly developed theories about the formation of the solar system and from studying the reflected sunlight from their surfaces. At the present time, we have no manned spacecraft that could reach an asteroid. What we do have, however, is our wonderful selection of robot vehicles. They will be our first prospectors. Fortunately for the future of our Martians, they will soon be set to work.

The NASA unmanned spacecraft *Galileo*, outward bound for Jupiter after a launch in 1986, will give us our first close look at an asteroid. The side trip will not, at about $20 million, be cheap; but that's a tiny fraction of what a special mission would cost. If the slight diversion on *Galileo*'s trip to Jupiter works out well, the cameras and other scientific instruments aboard will show the surface and composition of "29 Amphitrite," as the asteroid is called. The European Space Agency will launch Agora, an acronym for Asteroid, Gravity, Optical, and Radar Analysis, in the early 1990s; it will explore Vesta and possibly two other bodies. There is little question that in the late 1990s we will begin asteroid mining. As with anything else, when there's a demand, there's a supplier.

As space industrialization grows, so will the demand for raw materials that are not expensively shipped up from Earth. There will come a point, and it may not be very far in the future, when mining the asteroids instead of the relatively poor ores of the moon will be regarded as practical and part of the evening news. That will be the time for the new Martians to step in. The main body of the asteroid belt is far closer to Mars than it will be to Earth or its growing space industries.

Robot spacecraft operated from Deimos would scan the belt for likely candidates. Computers would make lists of the millions of tons of materials required in five years, in the same way that today we can call up the world's supply of oil reserves, gold reserves, or the stored total of almost any metal or raw material. Once the robots had done their job, the

Martians would send out their prospecting ships. On location, the miners would launch a solar mirror into orbit around the body; the mirror would pour heat onto the surface dust of the little world. From their cargo holds, the Martians would drop tractors onto the cratered terrain. The mining tractors would chew up minerals at more than a thousand tons a day.

If the asteroid were of the common variety, its surface would be rich in carbon and water, which would be heat-driven from the surface by the mirrors. Iron and nickel would be removed by magnets. Part of the asteroid would be used as fuel for a mass driver.

Other useful metals would be separated from the body by a process called "gaseous carbonyl (Mond)." There are no moving parts in this incredible technology, used on Earth today by International Nickel Company, and its energy and heat requirements are fairly low. Using the process, our Martian miners would pass carbon monoxide over the metal-rich ore from the asteroid. Under the right temperatures and pressures, the metals would form gaseous carbonyls, which would be sent to a collecting chamber. Inside the chamber the metals would again be subjected to specific temperatures and pressures; the result would be deposits of nearly pure metals in the chamber. Different metals would require different temperatures and pressures.

Upon completing the mining operations on an asteroid weighing two million tons, what the Martians would have is half a million tons of water-ice, carbon, iron, nickel, and nearly pure metals of the more expensive kind. They would leave behind a rather battered hulk of a million and a half tons, mining operations in space being no less likely to be unsightly than they are on Earth.

The results of Martian industriousness would take four years to arrive near Earth, using a complex but entirely possible series of gravity brakes. The "pile" would swing by the moon and then Earth, and as a final maneuver it would swing back by the moon again, after which it would be captured by Earth's gravitation and remain in an orbit nearby.

On an accountant's ledger, the return from the asteroid mining operation would gross somewhere between 50:1 and 100:1, meaning that even after the expensive expenses, asteroid mining by the new Martians would be a very real way the colony could pay for the tremendously costly building of the first city on Mars, the space elevators, the mass drivers on Deimos, the mining vehicles, and most of all, and most important to them, the changing of Mars to a world more like Earth.

9 THE RED EARTH

THEY SAY ALL BABIES are born with blue eyes, even those who are later the darkest-eyed. Perhaps this is not quite universally true, because it was not true of Deirdre Channon O'Neil. Born she was with green eyes, and they stayed that color. Not an emerald green or a bottle green, but a kind of deep, mossy gray-green shot with tiny flecks of gold. There were those who said her eyes were a genetic fluke transmitted by a distant grandfather; there were others who said, somewhat privately, that it was the first clear evidence of human genetic mutation caused by cosmic rays and space travel.

Never mind. They complimented her hair, which was thin-stranded and rich auburn and which, in adulthood, fell in long, deep waves below her shoulders. In zero-g, the hair wound around her neck from one side to the other, depending on the direction of her own motion through the air as she "swam" through the hub corridors of the space station. At times, when she was coming down from the local vertical to what we would call a deck, it fanned out above her head, a thousand-filamented spray. She looked for all the world as if she were in a swimming pool, sinking.

There were some who said she was beautiful in an alien sort of way and they always emphasized the alienness. For one, her forehead was broad, the brows fairly thick, and her cheekbones were rather large and prominent. Had she been more moon-faced, she would have looked Oriental, except for the limpid green eyes.

Perhaps the alienness was intentional; she knew from the earliest age that she was different from other girls. True, she was, by reason of birth; but by the time she was six or seven, she was not the only child of space for the press to fawn over. She was well into her teens when the first child was born at the Mars base.

No, it was a real alienness about her that struck all those who met her. She had a look (many found it hard to describe) that came from deep within her; it was not so much that her eyes penetrated those with whom she was talking, but they locked on in a kind of computer-like way and didn't let go.

She was also curiously built, leggy like a dancer with thin legs from her hips and a skeletal structure to her shoulders that was wider and somehow faintly distorted. If a dwarf can be said to be a compact and mildly distorted form of human, she was the opposite. The doctors thought it was caused by the ten months she spent in zero-g on the return to Earth and the years she spent in the space station. They were probably right, but *she* thought it was because somewhere inside she was a Martian. When she realized she should never live on Earth — the slight heart murmur was real and indeed dangerous — she examined her other side, her feelings about the Red Planet, and thus arrived there as the first lady of Mars in her twenties.

Now she is standing beneath the high dome of the Outland Station of Mars City and watching the dirigibles lift their precious cargo of genetically modified blue-green algae into the darkening sky. A storm, predicted to be a very mild one, will sweep through this part of Mars; it is a slow-moving front, and the dirigibles will return long before it hits.

The auburn hair is streaked with gray; it still falls below her shoulders. The strange eyes, still green and flecked with gold, stare at the panorama outside the dome as the ships lift off one by one. Her body looks less distorted on Mars, as if the Martian gravity put back something that was taken away once. Even by Martian standards, she is growing old.

Her son is by her side, looking alternately at her and at the departing ships. He has her eyes, the wonderful gray-green, but without the flecks of gold. He is lanky, with his weight filling out a slightly stretched frame. His hair is darker than hers, but the faint vibration of red is in it yet. Had they not been together, had they been seen separately on different days, someone would still have remarked that they were connected.

He would like to pilot one of those ships, he says, or be an asteroid miner. Something adventurous and exciting, anyway. He would like to be an explorer like his father, now long encased in the Martian sands he so loved and in which he was lost during a planet-wide dust storm.

The algae runs are a final element in changing the face and soul of Mars. The Martians have been planting algae for years in the deep gulleys and arroyos of the planet. They have been experimenting with patches here and there, little lacework fields of green, gray-green, and blue-green. Now it is time for the massive seeding of large areas so the atmosphere will finally come around to one in which Deirdre's son, or perhaps her grandson, can stand one day on a Martian plain and feel the dusty winds in his hair and on his unprotected face.

The event may be anticlimactic, as so many events are. The twenty-first century will be anticlimactic to those of us in the twentieth, when it finally arrives. We are all human with the same feelings; time goes so slowly and then too quickly. The twenty-first century was a distant eighty-five years away, four generations, when the world erupted into the Great War. When nuclear energy was first unleashed the year 2001 seemed a very distant half century away.

Now the anticipation has almost turned to fact. It is near the end. We are only a few years from the century's end. Most of us will see it change from the old one to the new. We made it. When we read in a newspaper or a book that something will happen "by the end of this century," we are no longer quite as ready to fling the paper or book in the corner and utter, Who cares? Suddenly there will come a day when we wake to the year 2000.

Terraforming will be a little like that for the Martians. It will seem to them at first that terraforming the atmosphere will be forever in their future, that they will never see the sensitive barometric instruments of the scientific stations creep slowly upward, indicating that the air is becoming more dense, the greenhouse effect is growing, the planet is warming, water will flow once again in the creeks and rivers that are the channel lands.

Perhaps there will be a giant barometer in the hub of Mars City, like a great railway clock in a Victorian train station. The travelers waiting for their little subway trains will glance at it from time to time; they will watch for any slight movement, because each infinitesimal click of the needle up the notched scale of the dial is one more step toward a Mars where their children or their grandchildren can run about the surface and play and do those other things their ancestors, not terribly far removed at this date, did on the distant Earth.

Simply raising the atmospheric pressure of Mars won't make the planet a place for mankind to live without environment suits. Melting the ice at the poles, throwing water vapor into the air, releasing water from the permafrost regions — that is only the beginning. The air, no matter how dense, the winds, no matter how warm, are not our idea of a beautiful day under the strange sky. We can't breathe that stuff Mars has for air, no matter how dense it is or how warm the world becomes. Nevertheless, changing an impossible mixture of air can be done in the same way Earth's air was modified. Billions of years ago, the atmosphere of Earth, called the primordial atmosphere, had no oxygen.

The early atmosphere was subjected to lightning, intense solar ultraviolet radiation, cosmic rays, and volcanic heat; the incoming energy stimulated the molecules of the atmospheric gases to form five new compounds. The compounds became condensing droplets of matter that fell into the oceans, which were a primordial soup. The compounds that fell were the chemical letters of the genetic alphabet: adenine, thymine, guanine, and cytosine. Together, they form deoxyribonucleic acid, otherwise known as DNA. The fifth letter of the genetic alphabet formed in the primordial atmosphere of earth was uracil, which is a component of RNA, the "yes man" of the DNA molecule, the genetic messenger that carries out the orders.

The production of the five basic genetic ingredients from atmospheric gases surrounding a primordial soup has been done in the laboratory: in 1983 the process was duplicated at the University of Maryland's Laboratory of Chemical Evolution. The five bases have also been found aboard a meteorite that dates back to the formation of the solar system

4.5 billion years ago. The same building blocks of life occur in fifty-seven different molecules discovered in the universe. The conclusion is that life is a relatively common occurrence, if the conditions are reasonable enough.

The alphabet soup of the primordial earth produced more complex forms, "tending," as the biologists say, "in the direction of life." The biological component of the earth changed the atmosphere. One of these complex forms was a single-cell creature called a cyanophyte, which is very important to us and for the future of our Martians. The terraformed Martian air will still be 96+ percent carbon dioxide, a bit of nitrogen, and a trace of oxygen. We can use the same biological mechanism that occurred here so long ago to make the Martian air more like the atmosphere of our home planet. The process is called photosynthesis, and cyanophytes are capable of doing it.

Photosynthesis is the fixing of carbon dioxide and water into a starch by chlorophyll-containing organisms using sunlight. The by-product of this microorganic industriousness is oxygen. The organisms responsible on Earth are blue-green algae, cyanophytes. They were the dominant form of life — along with bacteria — for two billion years on this planet. They still live here.

The very earliest forms did not even have an oxygen component atom. We have found dim evidence of these single-cell creatures in the oldest rocks and have concluded they existed when Earth's atmosphere had no oxygen. Moreover, they didn't have DNA molecules in their nuclei to control chemistry and reproduction. They were truly *simple* life forms, not far removed from amino acids and chains of molecules wandering in the direction of life. But they were enough to do the job.

The growth and multiplication of the blue-green algae

was incredibly slow. The ultraviolet-ray bath on the primordial Earth was savage, the amount of oxygen available was nil, and the life form was inefficient. But it caught on fast by geologic time intervals and evolved into a more complex critter. About 1.5 billion years ago, nuclei appeared in the blue-green algae, and they assumed more or less the form we find them in today.

The new and improved cyanophytes did their job with spectacular efficiency, if not great speed. They were far more adept at converting carbon dioxide to oxygen through photosynthesis than their ancestors. In a billion generations, a billion billion generations, of these unthinking organic field-hands, Earth's atmosphere wound up 700 million years ago with 5 percent oxygen and much less carbon dioxide than it had originally.

There was enough oxygen at last for other forms to survive and multiply, and they did. In 100 million years, life on Earth changed from single cells to complex forms, multicellular organisms. Hard tissue developed as an anchor for muscles in the more evolved forms and the hard tissue left a fossil record that dates back 600 million years. Life on Earth was off and running.

The temperature of the "old," pre-terraformed Mars was barely in the survival range for lower forms of life, even cyanophytes. The ultraviolet radiation hitting the surface, almost unretarded by the thin air, was twice as great as most Earth life forms could tolerate. There was practically no oxygen, upon which all life on Earth, with one or two exceptions, depends. Cyanophytes are one of the exceptions. Life also needs nitrogen, sulfur, phosphorus, carbon, and hydrogen. Those, at least, Mars has, although in some instances not in the form required.

But upon a Mars to which a substantial greenhouse effect

has been previously introduced, the temperature, ultraviolet bath, and general planetary conditions are only hostile to humans. It is no longer correct to say that Mars could not support Earth life. It can, and the sorts of life that can live there are exactly the kind that would be most useful to us: blue-green algae.

To change Mars by biological means alone into a planet where mankind (or much else) could live is impractical. It would be nice, of course, to contemplate doing the whole job from a distance and at vastly reduced expense: robot space-craft could be sent carrying biological terraformers. There would be no need for a manned landing, no grand design for an underground permanent base, no need for giant space mirrors. We could just sit back and let the robots and the little gas gobblers do the work. When it was done and Mars was more to our liking, we would land there and step out of the spaceship into a warm, sunny summer day and inhale a deep breath of pure Martian air.

The problem with that is simple: it would take too long. The old Mars was so marginal that photosynthesis would take place only three to four hours a day, no matter how hardy an Earth microorganism was introduced. If a quarter of Mars was covered with the little creatures, the same kind of blue-green algae that were primarily responsible for chang-ing the atmosphere of Earth to what it is now, only 5 millibars' worth of oxygen would be generated in 7,000 years. To pro-vide enough for human breathing — 100 millibars — could take 50,000 to 150,000 years. This is why the Martians will be experimenting in deep basins and in their greenhouses for some time after their base is established instead of running around strewing cyanophytes everywhere. It wouldn't do much good yet. The Martians will have to wait until the conditions on the planet have improved to the point where the algae can work quickly.

If cyanophytes were seeded on a warm Mars, one with a reasonable quantity of liquid water and water vapor in the atmosphere, they might grow and reproduce happily. In their eternal life processes of birth and death, with a lifetime measured in intervals tiny compared to ours, they would give off oxygen. The oxygen would go into the Martian air, where it would one day be useful for us. It's the timetable that is the problem. We don't want to wait even a few thousand years for the little microorganisms to complete their endlessly mindless task.

What we want to do is improve the cyanophytes, if that is possible, make them more of a CO_2 gobbler than they are now. There are several ways this could be done; one of them involves the space shuttle and the final years of the twentieth century. We can start preparing the biological terraforming agents for Mars *now*.

In November 1983, the space shuttle *Columbia* blasted off with the European Space Agency's *Space Lab* aboard. This billion-dollar lab has a variety of uses, all of which will contribute to the conquest of space in the twenty-first century. Of particular interest for the future of Mars was a series of experiments involving the biological packages aboard the lab: West German scientists sent a load of *Bacillus subtilis* off into space.

Bacteria are among the hardiest of Earth's life forms. Bacterial spores have been "revived" after being dead for more than 7,000 years in a lake bed. If it weren't for the miracles of modern medicine, the bacteria would win more often than not in any contest with humans, which are a pretty hardy and adaptable lot too.

The scientists who sent *Bacillus* into space wanted to find out what effect zero-g, a vacuum, and intense solar radiation unretarded by an atmosphere might have on the bacteria. The scientists were most interested in the survival rate, what mutations came about, and the repair rate of the DNA.

Among other interesting facts coming from the experiment was the discovery that the mutation rate was *ten times* what it is on Earth. It suggests that the laboratories aboard the space station and other manned structures in the future may be able to produce some cyanophytes that will mutate to something more useful for us on Mars. Cyanophytes in zero-g and strong ultraviolet radiation might rapidly evolve into a superman sort of little creature that could produce oxygen in Martian sunlight conditions. Not only would it be more useful as a terraforming bug, but the efficiency with which it converted sunlight to oxygen might be many times what it is now.

But shipping ten tons of blue-green algae off into space to mutate into something better may be going about the problem the hard way. There's rDNA, technology's finest and darkest side, depending on who is doing the evaluating and what the end product is intended to do. For more than a decade, the development of recombinant-DNA technologies has given us the ability to control the basic processes of life. We can isolate DNA, which is a long molecule from a living cell, from genetic material and manipulate it. We use biological glue called enzymes to "cut and paste" special DNA molecules onto the original ones. The result is new life by any other name.

If the scientist doing the work is one of the good guys, then the new life may be beneficial to mankind: more effective drugs, greater crop production, crop strains that resist disease and don't need fertilizer. We can make biological garbage collectors who like toxic waste, whereas none of nature's creations do. If the scientist doing the work is a bad guy, the imagination freezes immediately. New and highly effective forms of fantastically dangerous biological warfare agents are possible — the *Andromeda Strain* problem.

The research into gene splicing is far advanced; it is no longer a paper technology. The first patent for a genetic engineering process was issued in 1980; the second, a "product" patent, was issued in late 1984. Human insulin made by genetic engineering techniques (called Humulin) is already on sale by Lilly and Company.

Since we are several decades away from establishing any sort of permanent base on Mars, there is plenty of time to develop a special form of cyanophyte that would eat up the carbon dioxide in the Martian air and give us volumes of oxygen. The laboratory techniques necessary are available now, and sexual recombination and DNA transformation have already been successfully tried on blue-green algae. All we need is the firm resolve to begin the work.

Not only is it possible to use sexual recombination and DNA transformation to produce more efficient cyanophytes, but there is the chance that *synthetic* genes could be cut and pasted onto the gene structure of cyanophytes, resulting in a Martian gas gobbler of unprecedented efficiency. There is no single species of cyanophyte that has the characteristics we require for Mars, but it would be quite possible to build one with synthetic genes. If highly sophisticated rDNA-modified blue-green algae are developed for our use, then the question of introducing a breathable atmosphere to the planet may be solved. So, depending on the degree of modification of the little creatures, is the time frame. The super-cyanophytes could do the job in less than a hundred years.

Terraforming on a massive scale can get the air pressure

up to near Earth's. Blue-green algae can provide an oxygen component in a reasonable interval, we think. But what we get then is air that is CO_2 and oxygen, whereas the air of Earth is nitrogen and oxygen. On Earth, our CO_2 component is a miserable 0.3 percent.

Could we breathe it, that mixture of CO_2 and oxygen that we will have worked so hard to achieve? The answer is ... maybe. Actually human beings don't *have* to breathe nitrogen with our oxygen, we just prefer it that way. But a mixture of oxygen and any of the inert gases does fine for us. Nitrogen is an inert gas, of course; so is helium. And so is CO_2. Deep-sea divers often use a helium-oxygen mixture, and it has no particularly harmful effects except that it makes them sound like Donald Duck since our vocal cords were not designed by nature to work well in a helium-based atmosphere. So there is a chance that we could breathe the terraformed Martian air, but at present there is considerable doubt that it would be a particularly good idea.

There are reasons other than convenience for having nitrogen. The whole of Earth's delicately balanced ecology is based on nitrogen, and unless we want to tinker genetically with nearly everything we send to the Red Planet — except for cyanophytes and lichens — we eventually need more nitrogen.

The Martian air is 3 percent nitrogen, but various estimates of planetary outgassing indicate there should be vastly more nitrogen on Mars than we can find in the atmosphere. On Earth the nitrogen gas wound up in the air through biological means; our carbon dioxide went into the oceans and rocks. Since we don't think there is any life on Mars, we are not surprised to find so little nitrogen in the air. But exactly where *is* the missing nitrogen? Or are the estimates based on planetary evolution wrong? After considerable study of the data from the best source available to us — the *Viking* explorations of 1976–82 — many scientists have reached a tentative conclusion about the "nitrogen problem": the gas is bound up in the soil, presumably as nitrates, nitrites, or possibly ammonia compounds.

If the nitrogen is in the soil, it can be put into the atmosphere by biological means through the use of nitrogen-eating bacteria that return nitrogen as a gas to the air. There are plenty of these available on Earth for us to modify and send to Mars. Some of these bacteria convert nitrites and nitrates into nitrogen gas here; they would work just as well on Mars, especially if we tinkered with them through rDNA. Ammonia is gobbled by other bacteria and turned into nitrites, which in turn become nitrogen gas through more bacterial action.

The greatest problem for the terraformed atmosphere might be caused by the tilt of Mars on its axis. In ten thousand years or so, the planet might go through one of its periodic ice ages again, which are caused by tilt variations. The consequences for the Earth-like air might be disastrous: it would freeze out. There have been some fancy suggestions on what to do about this problem if and when it occurs. One has been to introduce a gravitational counterbalance that would stop the Martian axial tilt from changing around.

There are several kinds of gravitational counterbalances that would work. A reasonably big asteroid could be moved in from the main asteroid belt and shoved into a reverse-direction (retrograde) orbit around the planet. Neither of the Martian moons could do the job; they are so small that they would have to be moved in too close to the planet's surface. Phobos, if it were close enough to fix the tilt, would be rapidly ripped by the gravitation of Mars from the orbit established.

The details of the solution will have to be worked out

by some future engineering genius. At any rate, the Martians probably aren't going to be worried about strange possibilities that may arise ten millennia from now, so the question of what to do may not come up. It has also been suggested by many geologists who have studied the data from *Viking* spacecraft religiously that no new major natural climatic changes are likely; they think Mars has reached some kind of equilibrium that will last for a long time.

If the air pressure were changed and the atmosphere had oxygen, and nitrogen, what sort of a world would Mars be? There is a tendency to think that it would become a sister Earth, a slightly smaller and poorer sister. This is a very narrow view, and it's basically incorrect. Mars will never be much like Earth, except in gross terms, no matter what we do to it and no matter what fantastic engineering schemes or biological experiments we perform. It will be a planet where we can live reasonably comfortably as a human species, but Mars will always be Mars. That, after all, will be the ultimate fascination.

Clouds on our new home will be much higher in the sky — about three times as high. There might be a bit more ultraviolet radiation than we are used to, because a proper ozone layer might not form. If one does, it is not yet clear if it would remain in place. The ozone layer is what protects Earth from the more savage ultraviolet radiations of the sun and what protects us from terrible sunburns.

The sky will not be quite so salmon-pink. It is that color now because of the high concentration of dust particles in the air, and Martian dust particles are red. If it begins to rain (and it would when the atmosphere became sufficiently terraformed), the dust particles would wash out, leaving the sky a very Earth-like blue with perhaps a touch of the red color remaining.

The great channel systems of Mars that once held running water will fill again when the atmospheric pressure reaches a certain level and the planet is warmer. There will be rivers, but not exactly what we're used to. They will be big enough (some of the Martian systems will rival the Amazon) but we may not want to go rafting on one or take a swim. For one thing, there may be ice rivers. Mars will not be fundamentally warm enough for year-round water flow for many years. On a still-cold Mars, any river water would quickly freeze, producing a surface layer of ice. The result would be ice-sheathed rivers flowing for hundreds of miles. Because the ice cover would trap air (thus raising the pressure), the water beneath would remain liquid.

The rivers of Mars might very closely resemble the Lena Delta in northern Siberia; the waters and tributaries there flow to the northern Arctic Ocean. In winter they freeze over, but the liquid underneath still keeps onward in its journey toward the pole. The surface ice is thick enough most of the winter to walk on, despite the relatively warmer water flowing just beneath.

On Mars, the water under the ice sheath would occasionally break out and cause minor flooding. This happens in the Lena Delta, too. The broken ice would jam at sharp turns of the river and at islands. The resulting dam would halt the flow until the pressure behind it forced the flowing liquid through in a spectacular burst of jumbled ice and rushing flood waters, causing damage to the river channel.

Flooding waters tend to take with them whatever is around. There are of course no trees, bushes, houses, or anything like that for a Martian river to sweep down to the nonexistent (for now) Martian sea. Like rivers on Earth, however, the Martian river will carry along tons upon tons of soil; it will also carry along the minerals in the soil, and some of those minerals will dissolve in the water. We describe the mineral content of water in many ways, but the most common expression is "hard" or "soft." The stuff on Mars will be harder, as the old expression goes, than the hubs of hell, because of the metals, sulfates, and calcium carbonate in the soil. The rivers may be a little on the salty side, too.

Always the Martian view will be strange to our Earth-born eyes. Mars is a smaller planet than ours, and the horizon will seem much closer; the landscape, under certain lights, will appear foreshortened. A similar distortion occurs in movie scenes that are shot with telephoto lenses; what we recognize as distant objects seem very close. The effect is particularly pronounced in shots of long, undulating country roads.

There will always be the strange, hurried cycle of the tiny moons as they go about the sky in their odd orbits. There will always be the 38 percent Earth gravity, which will make us all feel as strong as weightlifters. Television waves won't travel as far; they are line-of-sight transmission and the planet is smaller. The same is true for radio waves, unless the atmosphere of Mars has a well-developed ionosphere. None of this will have much meaning, since all radio and television will probably go through satellites.

The calendar will be odd, but we have become used to several different calendars during our stay on planet Earth; we are immensely adaptable in the ways we count our time left. The seasons will seem a bit long to us — roughly twice as long — but we can adapt to that too.

The terrain will never be quite Earth-like. On Earth there are two basic kinds of terrain: the relatively upraised continents and the more extensive ocean basins. Mars also has two distinctive forms, but they are unlike ours. Most of the southern hemisphere and the equatorial regions are highly cratered plateaus and hilly terrain. This area contains maybe 100,000 craters of all shapes and sizes, most of which date from almost four billion years ago, near the time of the formation of the solar system, when all the planets were bombarded with cosmic debris. Most of the northern hemisphere is a series of plains, generally lower, and with fewer craters.

The largest Martian impact basin is Hellas. It is more than 900 miles in diameter and lies halfway between the equator and the southern pole. Generally the craters are smaller than those on the moon, and although Mars has not had nearly as much erosion as Earth, the Martian craters are more eroded, and less well defined, than those on the lunar surface. They are also unique in some places, with layered slopes; these rampart craters occur all over the planet.

The massive volcanoes tower far higher than anything on Earth: Olympus Mons is at least 16 miles high. Mount Everest is less than 6 miles high the way we count things here. Heights of objects on Mars are a bit arbitrary, since there is of course no sea level as a reference point. The reference point generally favored relates to the atmosphere and is called the "6-millibar level," corresponding to that atmospheric pressure. Some of the lowlands have a 10-mb pressure; the top of Olympus Mons has a much lower pressure. When the planet has been terraformed and the atmosphere changes, this system of measurement won't work, so something else will have to be found. What might be used is a reference to a real sea level if things go right for the Martians.

Olympus Mons's magnificent proportions are caused by the lack of plate tectonics. If Olympus were on Earth, the movements of the plates would have cut it off from its magma source before it got much larger than the volcanoes of Hawaii. On Mars, it just grew and grew. The caldera on the summit is nearly 50 miles across and the great escarpment that surrounds the gigantic base of the volcano is more than 3 miles high.

We are not quite sure whether Olympus Mons is extinct. When we first land on the planet, we may solve the mystery at last. There are certainly some weather patterns associated with the volcano that are suspicious, and there have occasionally been reports — most of them made long before spacecraft visited — of unexplained bright flashes in the area of the volcano. While scientists tend to mistrust and discount much of the telescopic observation of Mars (the canals being a good reason), it is also possible that astronomers using telescopes on Earth have in fact seen Olympus Mons in a rare eruption. If they have not seen an eruption, then they may have at least seen evidence that the volcano is not entirely extinct.

There are three other great volcanoes that will dominate the Martian sky to anyone within seeing distance. They form a line running from northeast to southwest, about 600 miles from Olympus. They are very young features of the planet, though they thundered and roared for hundreds of millions of years. The other major volcanic region is in Elysium, where three more volcanoes perch on another planetary bulge. Elysium Mons, the largest in this group, is more than 7 miles high and more than 150 miles across the base.

Mars has a feature that appears on no other planet in the solar system: a low, broad, central-vent volcanic structure that geologists call a *patera*, which means saucer. The largest one, Alba Patera, is 1,000 miles in diameter but stands only a few miles above the Martian surface. There is a ring of jumbled terrain and fractured surface that surrounds the saucer for another 500 or 600 miles.

On Mars there are great expanses of lava not unlike those of Earth; the Deccan lava plain in India covers thousands of square miles, as does the one in the northwestern United States near the Columbia River. The most extensive of the Martian lava plains are in the north, quite distant from the cratered highlands of the south. Chryse Planitia, where the old *Viking* landed in 1976, is one of the Martian lava plains.

Other than the giant volcanoes dominating the landscape, the single most distinguishing feature of our favorite planet is the canyonlands. There is a Great Equatorial Canyon Region that dwarfs anything Earth — or any other planet, as far as we know — has ever produced. Mariner Canyon, also known as Valles Marineris, is 3,000 miles long and varies from 80 miles to 400 miles wide. Some of the deepest sections extend more than 5 miles down, and Valles Marineris isn't the only major canyon; east of Noctis Labyrinthus are the long, sinuous, narrow trenches of Tithonium Chasma and Ius Chasma; enormous landslides have occurred in both, and Ius has a series of tributaries stretching like centipede feet from the narrow body of the canyon.

The great channel system of the planet is not the long-lost Martian canals, at least not in the sense of having been built by a civilization of intelligent Martians — but they probably did once carry water. Some of the channels are more like valleys or troughs, but a significant portion of them look as if they have had running water in them. Many of the broadest occur around Chryse Planitia and run from the heavily cratered areas out onto the flat plains.

The narrower channels without tributaries meander along looking exactly like mountain streams on Earth except that they have no water in them now. Another kind of channel network runs down the sides of craters and goes out onto the plains. There are also large volcanic grooves in the volcano district that look like those carved by lava flows on earth.

Terraformed, Mars will resemble the more inhospitable parts of Earth: Death Valley, the bleakest parts of Utah and Arizona, the Sahara, and most of Mediterranean Africa. But it may not always remain so: someday it might have a shallow ocean or two.

The great channels will carry water again when the greenhouse effect takes hold. Even those that originally did not carry water will, because of their shape; they will drain water from the highlands to the lowlands. The crater arroyos will carry liquid to the surrounding areas if the crater basins fill up with rain. And when the waters of Mars begin to flow again, directed to some extent by the channel system, they will flow *to* somewhere, they will collect somewhere. It is possible that they will collect in what will someday become a small Martian ocean, perhaps even two oceans.

English is a harsh language. It is neither as precise as German, nor as descriptive as French, nor as poetic as Spanish. Take Desemboque, which is a little place on the west coast of Mexico, two-thirds of the way up the Gulf of California. It is a few miles from the confluence of two rivers, the Coyote and the Magdalena. *Desemboque* means "the place where the rivers meet the sea." It's a word that rolls from the tongue more gracefully than *delta*.

On Mars, Desemboque will be the place where the runoff of the rapidly warming permafrost will go. There will be many deltas on the Red Planet before this terraforming is through. But the question of whether the deltas will spill into a real sea is inextricably bound up with the issue of just how much water there is. But does Mars even need an ocean? Are we simply trying to make it more Earth-like?

Earth has plenty of ocean, enough, in fact, to cover our planet to a depth of more than a mile, if the oceans were spread evenly over the globe. That is a massive amount of water, far more than Mars will ever have, and it plays a significant role in the health of our world. By comparison, the water contained in our atmosphere as water vapor is tiny: only enough to cover our world with slightly more than an inch of water.

Oceans on Earth play a very important role in the cycle of everything, but the two most important roles are as the ultimate source for rainwater and as a temperature stabilizer. They profoundly affect the climates of coastal areas, and winds over ocean areas cause evaporation which brings rain.

Fortunately, it is only the top 300 feet of the ocean that has any major effect on heat transference around the globe and on weather patterns, including rainfall. Thus if an ocean is absolutely necessary for the stability of an Earth-like climate, we can probably get away with a *shallow* ocean. That's lucky for the Martians, since the water supplies may be limited. No one doubts there is a great deal of water on Mars, but there is considerable bickering on exactly how much. One of the reasons why NASA wanted a *Mars Geochemical Orbiter* to circle the planet in the early 1990s was to determine, if possible, the extent of water supplies.

Most assumptions by scientists now, however, lead to the conclusion that Mars has enough to make a small ocean. Even if it did not, to a world of the future where terraforming Mars is a gigantic but feasible engineering feat, it might not be a problem. More water could be imported from the asteroid belt from a water-ice asteroid or perhaps from one of Jupiter's moons, one that we know is composed of water-ice.

The asteroid would be brought from the main belt and left in an orbit quite distant from Mars. The Martian miners would then break off large chunks of ice and send them down to the planet below on a precisely engineered trajectory. The resulting impact would create the first new crater of any size in millions of years. It would also provide the planet with a sizable water supply, depending on the water-richness and size of the asteroid chosen.

Where the oceans of Mars might lie is reasonably simple to guess. The warming of the planet, the changing of the atmosphere, and the return of running water won't change the terrain in gross geological terms. Mars will still have highlands and lowlands. Since liquid on Mars flows the same as it does anywhere else in the solar system, including Earth, Martian water will fill up the lowlands, leaving the highlands as "continents."

If enough water were released, oceans could cover as much as 35 percent of the planet, although they would be quite shallow. They would also not have serious tides. The tides on Earth are caused by the gravitational action of the moon, which is almost a quarter the size of our planet, and to a lesser extent by the pull of the sun. The gravitational action of the little Martian moons on a Martian ocean would be practically nil by any standards we are used to; the only tidal action would be caused by the distant sun.

If oceans were finally introduced, either as the result of importing an asteroid or if there are more reserves of ice than we currently think, the climate pattern would be somewhat like Earth's. That is, there would be the *kinds* of weather patterns we are used to (wind, rain, thunderstorms), but the effect might be a bit different. Hurricanes might come screaming out of the narrow ocean areas to lash the land. Rain on the coastal regions might be very high, much higher than on Earth. Monsoons might be more common than gentle afternoon rains. With sufficient moisture in the air, the slopes of the great Tharsis volcanoes might become glaciers. The northern polar area might be a great sea, a cold body of water with rafted ice hummocks in winter — a shallow duplicate of our Arctic Ocean.

There is one interesting possibility in the future of Mars that has a strange irony. If oceans form, they may do so as small, isolated bodies of water. In the Hellas Basin, for instance, a sea 800 miles in diameter could exist. Like most of the basin areas on Mars, the Hellas Basin has channels running from it. When the basin had filled up, massive flooding would cause near-catastrophic erosion down the channels as the waters of the basin tried to find an outlet to any nearby larger (and lower) body of water. Even the partial emptying of a large sea into another can profoundly affect any climate — it has done so in Earth's past — and the Martians might want to avoid massive changes in their delicate balance caused by small oceans trying to move around and become big ones. One way to stabilize the system would be to connect all of the large bodies of water with canals. It would come full circle, then: the Martian canals would have been built by a race of intelligent beings, just as Lowell said in 1895.

There is one other major factor that may be a problem for our future colonists. Mars has a weak magnetic field, as far as we can determine from the investigations by the Viking mission. The role of Earth's quite strong magnetic field is poorly understood, but there is evidence that disturbances in the field, especially a reversal, can be very harmful to life on the planet.

At several periods in Earth's past, some species in the oceans were wiped out by variations in our magnetic field. When the field reverses, it is weakened by a substantial amount; some scientists believe that it becomes zero for a prolonged interval. It is unclear whether the harmful effects on living creatures are due to the reversal or to the weakening.

Laboratory tests have shown that many living species show definite effects from a weakened magnetic field. Bacteria lose interest in sex. Some ocean life shows restricted ability to move. Birds exposed to a reduced magnetic field act as if their wings have been clipped. Mice die more quickly and they too lose interest in sex.

By the time we get to Mars, we may have sufficient information about the biological effects of changes in magnetic fields to tell us whether we need to do anything about Mars's field. Experiments aboard the space shuttles and in laboratories on a space station may provide the answer long before our terraforming is under way.

Based on what little we know now, there may be one interesting solution. It is a solution that would have some nice aesthetic side effects. Of the rocky planets, only Earth has a sizable moon. Venus, of course, has no moon; neither does Mercury. The moons of Mars hardly count. The absence of moons has been apparent for hundreds of years, of course, through observations with telescopes. What we did not know until spacecraft visited the planets was that only Earth has a very strong magnetic field.

It is likely that Earth has an exceptionally strong magnetic field partly because of its moon. If so, then we may have to think about giving Mars a sizable moon in the future so that the delicately balanced life that we're implanting there will have at least something like the same chance for survival as it has had on Earth. If we can talk about putting oceans on Mars, or having them come about as a normal consequence of our terraforming, we might as well consider putting up a decent moon. Neither of the tiny moons of Mars is worth much, and by the time we have terraformed the planet we may have shifted half the surface dirt of both of them down to the poles. To a future society that can contemplate changing an entire world, moving in a satellite isn't much of a problem. Any of the smaller moons of Jupiter would serve.

To make the Mars–moon system seem more like the Earth–moon system, it might be nice to find a body about 1,000 miles across and put it at a distance so it would appear about the same size as our moon does in the night sky.

10 AND SO ON AD INFINITUM

THERE ARE PEOPLE who think that if Mars is ever reached, it will be a voyage of exploration only. They cite the extreme costs of even the most well-planned and cost-effective manned mission. As far as colonization goes, they shrug their shoulders and talk about the severe Martian climate, the horrible atmosphere, and the lack of viable self-sufficiency without massive doses of high technology sent from Earth.

There are others who are very optimistic about colonies on the planet but think they will forever be restricted to underground warrens. Among the latter are many who believe that the Martians will be vegetarians and that the biology of Mars will be confined to the human colonists and assorted leguminous plants on the farm.

People who are persuaded the Martians will be strict vegetarians generally get into the efficiency of the matter; they tend to ignore human nature and human history. It is true that vegetable protein is converted into animal protein very inefficiently. Feeding animals — for example, horses, pigs, cows, sheep, and the like — and then eating them is a relatively frustrating proposition for anyone concerned with waste. The best conversion rate is only about 10 percent. Eating this way, 90 percent of the protein is lost, along with an equal proportion of other things we need for survival, such as vitamins. But upon a Mars that we have been so outrageously confident as to terraform, it seems unlikely that we would be satisfied forever with looking at beanstalks for lunch.

As a race, we not only aren't predisposed to being exclusively herbivorous; but we aren't very practical either. Practical people don't usually want to go homestead on Mars, they want to do something more . . . practical. Romanticism created Mars. Romanticism will populate it. Folks who are so unabashedly romantic as to establish a colony on a far-off and lonely little world that, for want of a better description, could only be described as one bad piece of cosmic farmland, may want a little steak with their potatoes. They may want a little company, too, preferably an agreeable kind. You can't take a cornstalk for a walk under a noonday sun filtered by the dome of the colony's farm.

And it is impossible to maintain an ecology involving only humans and algae. It's undignified and it would cause talk. What we want is a "closed-cycle" ecology — a real live, working biological planet, in other words.

In addition to the blue-green algae and the nitrate/nitrite-gobbling bacteria, a thousand species of "this and that" may be necessary to help a planet run smoothly, most importantly with stability and no danger to the human population. How to seed the right things and not the wrong things is a project that will tax the Martians' science and imagination.

Plague bacilli are large, elongated rods with rounded ends that stain at both poles in the parasitic stage. They are often shaped like safety pins. Rats, squirrels, rabbits, and 217 other rodents get infections from them. The infections are maintained by an insect vector, the ubiquitous flea. When the rodent–flea–rodent cycle touches the human race in any way, the result is bubonic plague.

The terrible *P. pestis* can survive for three months in dried mucus and five weeks in the residue from fleas. Jailed in a refrigerator, the bacilli have been kept for more than a quarter of a century. They still remain as violent and deadly as ever.

The virulence is legendary: a sudden onset, high temperature, rapid pulse, white coating on the tongue, nervous symptoms, fatigue, slurred speech, staggering gait. Penicillin is useless. The fever rises to 105°F, and death is hideous and painful. At full howl, bubonic plague unchecked by modern medicines has a death rate of six in nine.

The Black Death and its fateful residence on Earth is a minor lesson in what is called "planetary ecosynthesis." When Mars has been terraformed and there is oxygen, nitrogen, and reduced carbon dioxide in the atmosphere, when there is flowing water, moist soil, rain, clouds, it will still be a raw world. It must have a stable ecology if it is to be a second home for mankind.

We don't want to take *P. pestis* along to the Red Planet and there are quite a few others of Earth's microscopic inhabitants we won't want, either. The little virus that causes Venezuelan equine encephalitis is one. The Junin virus, responsible for Argentine hemorrhagic fever, is another (the mortality rate is double that of yellow fever). We don't want Machupo virus, Marburg virus, or Lassa fever virus, any one of which makes such things as smallpox and cholera look like the common cold.

The disease viruses are obviously things we don't want,

but what we *do* want may be much harder to determine. We need to know exactly what the role of an organism is in Earth's ecology before we can transplant it to a Mars that is sufficiently like Earth to accept it as a life form.

An interesting example of the extraordinary role of computers in this sort of problem comes from rocket-fuel research. Back in the good old days before computers, the hardy and near-suicidal lot who were rocket-fuel experimenters would get a hot idea over lunch, and armed with scribbles on a paper napkin, would pour two substances together to see if they would get us to the moon quickly. Sometimes all that went toward the moon were bits of the laboratory, often accompanied by bits of the scientists.

The rocket engineers were remarkably like Grand Prix drivers; there weren't many real ones in the world and the injury rate among the best and greatest was extremely high.

Fortunately for all of us, along came the computer, and it allowed rocket-fuel experimentation to take place within the hidden electronic paths of the hard and software. It was no longer necessary to take grave risks to investigate dangerous combinations looking for the sweet one. It all got very civilized and easy, and rocket-fuel engineers became a more sedate group of people.

Roughly the same thing will happen for planetary ecologists. No scientist is going to experiment with a real planet. The results might be much worse than mixing dangerous rocket fuels. It is simply too unpredictable and the result could be disaster. Future planetary ecologists will use computers to "introduce" various life forms to the Martian environment and see what effect they will have on the planetary ecosynthesis through computer models.

The computers will be able to predict accurately the net change in the planetary environment over any amount of time the scientists care to investigate. When a particular microorganism has been identified as necessary for the Martian ecology, it will be sent along to the planet. The organism, depending on its ecological function, may or may not be subjected to rDNA genetic tinkering to improve its efficiency. In most cases it is unlikely that rDNA will be used, since bacteria multiply prodigiously and they can therefore increase their net "efficiency" rather easily by common reproduction.

The rapid reproduction of bacteria is actually something of a problem, which is why computer models of planetary ecosynthesis are so necessary. If *E. coli* bacteria, the most common kind, were allowed to reproduce unchecked, it would equal the weight of Earth in a very short time. Humans carry large colonies of *E. coli* in their intestinal tracts, but the reproduction of the bacteria is controlled naturally.

Because of the inevitable dangers involved, all biological immigration to Mars is going to be heavily restricted and very carefully controlled. This may limit the introduction of microorganisms over time, but it does not necessarily mean that the introduction of plants and animals will be restricted in the same way.

It was always suspected, even when it was known that the conditions on Mars were extremely hostile to most Earth life, that lichens existed there. Lichens are a symbiotic organism of algae and fungus. The fungus partner offers protection from cold and provides, among other inorganic substances, water. Most fungi are sponge-like. The algae partner builds up organic substances and supplies oxygen via photosynthesis. Because of the character of the symbiosis, lichens can survive extremely cold and dry conditions. They grow on trees, walls, lava, and almost any exposed rock. On Earth, they are the dominant life form in the high Arctic areas and on high mountains.

Earl Nelson, the great-great-grandnephew of the British admiral of the same name, wrote in 1956 about life on Mars, naming lichens as the most likely explanation for the color changes on the Martian surface observed in telescopes with the coming of spring. Gerard de Vaucouleurs, writing in 1952, dismissed Mars as being unsuitable for almost all Earth life forms. The idea that lichens existed there, however, persisted long after the *Viking* spacecraft were launched in 1975.

The Viking mission didn't find any, and it now seems unlikely that there are indigenous Martian lichens. But as with most dreams of Mars, if lichens weren't there before, the conditions are so close that we can put them there now. We can certainly introduce them when the planet has been changed even a little from what it is now.

Lichens have any number of interesting properties, the nicest of which is that a few varieties require little or no oxygen to thrive. They are remarkably tolerant of a lack of water. That means they can be implanted on Mars at an early date in the colonization program, when the atmosphere will be very poor by Earth standards. Most of them take in CO_2 and give out oxygen, so they can double up with the cyano-phytes to terraform the atmosphere. (They are not, however, as efficient as the little blue-green algae; lichens' conversion rate of CO_2 to oxygen is perhaps a tenth as good.) They do have other qualities: they can grow almost anywhere; they are very hardy. They are also useful for holding the soil to-gether. A blanket of lichens on some regions of Mars would keep down the dust and also lay a friendly carpet of green or gray-green against the rusty red.

There are other algae than the blue-green kind that might be useful. A red algae called coralline can live in ocean depths of nearly 800 feet, an environment that has almost no light (0.0005 of the light at the ocean's surface). They are found on the steep sides of underwater mountains off San Salvador Island in the Bahamas. For all practical purposes, the little creatures can be said to grow in the dark. They would therefore be somewhat useful as a terraforming agent, since their efficiency at photosynthesis is one hundred times that of near-surface ocean algae.

Fine: Mars can be populated with blue-green and red algae and lichens; they may even thrive — certainly must thrive, if serious terraforming is to be successful. But growing even these rudimentary Earth forms may be extremely difficult on Mars. To grow more complex forms in the Martian soil will require a complex soil.

The Martian regolith is inorganic, as far as we know; that's why it is not technically correct to refer to it as "soil." Being inorganic means that there's no life in it, which means that there are no microorganisms. Thus no plants can grow in the soil, since nutrients are required by plants and the nutrients are cycled by microorganisms.

To grow plants in the Martian sands, we will need to supply the regolith with a suitable micropopulation. Once this has been done the soil should be reasonably suitable for a variety of Earth's flora, including the food-producing crops of the colony's greenhouses. Most of them will be familiar Earth varieties, although many of them may have been treated, before leaving Earth, to a little rDNA work. They will produce more, better, for longer, with less or perhaps no fertilizer, and they will be extremely resistant to any of the crop diseases that damage agriculture on Earth. Part of this will be due to successfully modified bacteria.

This sort of biological/botanical "improvement" has already been tried on Earth. A kind of rDNA bacteria produced in 1984 refuses, unlike its naturally evolved counterpart, to trigger ice formation in potato plants. The specially engineered bacteria are sprayed onto a row of potato plants and, as if by magic, the plants become highly resistant to frost damage. Any crop highly resistant to frost damage would be extremely useful on the surface of Mars at any time in the next century.

Once the regolith has become soil in places, plants other than the greenhouse crops may be tried. Some of Earth's ground covers would be reasonably successful on a terraformed Mars. Creeping Jenny might protect steep, rocky slopes; Scotch and Irish moss might add color to an otherwise barren landscape around the colony. If the soil is fertile enough, the full range of Earth's ground covers and colorful plants can be tried, from *Achillea,* also called woolly yarrow, an evergreen herb with narrow, feathery, gray-green leaves, to *Zoysia,* also called Korean grass.

You can't eat most flowers. You can't feed them to most animals. They are hard to take care of and they are delicate. They also attract bees, but bees may not necessarily be a part of the Martian ecology. Still, flowers are one of those most human of things. We must have them on Mars. Red roses for a Red Planet.

Bacteria are hardly much company. It is hard to ride seaweed. Flowers don't talk much. What will humans ship to the Red Planet to make life more bearable, useful, or pleasant?

If it could be said that astronomy was *the* science of the nineteenth century and physics was *the* science of the twentieth, then biology must surely be the top candidate for the next century. The last few decades have been only a glimpse of the future that may be only a short distance downstream.

The time scale for biological discoveries is about the same as for other sciences. It takes about fifty to sixty years to go from initial discovery to major social impact, about half that after intensive laboratory testing. The ubiquitous "pill"   .

is a perfect example: it resulted directly from investigation into steroid hormones which began around the time of World War I. By the 1930s the hormones' chemical structures were worked out and they could be produced artificially. This led to testing of artificial hormonal levels in human beings, a by-product of which was a useful artificial hormone that was an oral contraceptive.

Another incredible development in recent years has been the rDNA research that will be so useful when it comes to terraforming Mars and perhaps in populating it as well. Molecular biology in the last few years has been an explosion of knowledge, one that has shaken the foundations of traditional biology. At the center of the explosion is basic chemical information on what our cells use, store, and pass on to future generations. The understanding of the fundamental chemical processes of life has resulted in the predictable: manipulation of the chemical information. We now have the ability to re-structure the molecules that program living cells.

Genetic analysis of certain bizarre insects led us to iden-tify the genes that control the basic organization of the insect's body. This came from studying insects that looked as if they had just stepped off a Hollywood movie set under the macro lenses: fruit flies with legs growing out of their heads and things like that. Through study of the gene-connected devel-opmental control, it has been found that similar genes control similar functions in species from earthworms to human beings. Many animal species contain a stretch of DNA remarkably like that of the fruit fly. There is more than a casual rela-tionship between such diverse creatures as fruit flies, beetles, earthworms, frogs, chickens, mice — and humans.

Not only has microbiology provided a way to "improve" a species (a six-legged Martian dog is quite possible, but to what use it could be put is unclear), but advances in studying ordinary reproduction have made a shambles of old notions like Noah's Ark. For one thing, it is not necessary — nor desirable, in some instances — to have more than one sex of a species on the premises. Parthenogenesis, a twenty-dollar word for virgin birth, which is in itself perhaps a bit of a misnomer, was achieved in sea urchins in 1899 by Jacques Loeb.

Loeb's work was greeted with considerable derision among the scientific community, which tends toward conservatism more often than it should. The products of Loeb's laboratory were derisively referred to as "chemical citizens, the sons of Madame Sea-Urchin and Monsieur Chloride of Magnesium." Loeb's work led to attempted parthenogenesis in higher orders of creatures. The noble frog was next; it did not turn into a prince, but did dutifully produce a virgin-birth tadpole. This sort of research has led, in recent years, to experiments with frog eggs in which nonreproductive cells have been turned into reproductive cells. It became obvious that ordinary body cells (as opposed to reproductive cells) could be used to reproduce. The result was the biological *deus ex machina* called the clone.

In 1962 two healthy Border Leicester lambs were born to two South African Dorper ewes. This was the world's first instance of what is called "artificial inovulation." It's the reverse of artificial insemination. A female animal is encouraged to produce a large quantity of eggs through the injection of FSH, the hormone that is responsible for normally releasing eggs from the ovary. Then each of the eggs can be placed in the womb of another animal (a rabbit was the custodian of the eggs for the trip to South Africa in 1962) and the original "mother" can produce several hundred offspring a year instead of one. It can be done between different strains of animals: white rabbits can beget black ones, Friesian cows can give birth to Hereford calves.

If a calf is not exactly a calf anymore (as the example of the surprised Hereford above shows when it first regarded its Friesian mother), sex is not sex, either. The brain tissue of mammal embryos is fundamentally female. Maleness comes in when testosterone is produced on orders from the male chromosome. The effect of the testosterone can be altered in a number of ways, including neutralizing it with cyproterone acetate. Boys into girls, but not necessarily the reverse. The long-suffering rats upon which it was tried were "males" born with a vagina. After ovaries were implanted, they produced eggs and the rats behaved like female rats: *varium et mutabile semper femina.*

The use of rDNA-modified organisms exists in our own time. So do artificial insemination, artificial inovulation, some forms of cloning (a horse has been more or less cloned), and laboratory work toward changing sexes at will before and after birth. What the future will hold is certain to be more incredible as the explosion of biological discoveries increases. What *is* certain is that in their time the Martians will be able to choose what species they wish, in what size, in what sex, and with whatever degree of modification they desire. There will be none of this poking and shoving of noisy, ill-tempered animals into some future Martian cattle boat. Only lightweight molecular biology would need importation to begin a species, as long as laboratory birth techniques have been developed by the time Mars is colonized.

Want a hundred sheep? Send frozen reproductive cells in a plastic box off to Mars at no weight penalty for propulsion and taking up little room in the baggage car. When the herd-in-a-dish got to the planet, the biologists of the colony would fertilize some previously ordered female eggs and place the resulting mess in an artificial uterus. In a few months, there would be a lot of baaing around the place. If more sheep were wanted, more could be cloned, but it would be just as easy at that point to let nature take its course and have the sheep do the job themselves. Most sheep look as if they were cloned anyway.

The procedure works the same way for any of the other mammals. Imagine a Martian golden retriever, twice as intelligent as any breed today. The body would be like a cheetah's, all legs and hips, with speed to match. Barkless, the retriever would have near-human vocal cords that could convey a limited vocabulary. They would make fascinating pets for children, better watchdogs than anything we have now, and in a pinch they could double as babysitters.

Whales could be introduced into the oceans, and they would keep ocean life in check just as they do on Earth. Since the Martian oceans will be shallow and rather small, the whales could be made into a smaller species using rDNA techniques. You could get a smaller whale by the ordinary methods of selective breeding, but it isn't too practical: trying to wrestle with a mammal that outweighs us by a factor of several thousand and get it to breed our choice rather than its own is not easy. Big whales do pretty much what they want. If they notice people at all, they do so on about the level at which we notice a firefly. It's a physically distant relationship.

With the friendly dolphin, which is highly intelligent, we might do some engineering so that it will be capable of more complex thought processes as well as human speech. Experiments in dolphin linguistics have been in progress for some years now. Keeping track of a Martian ocean's ecology might take a massive scientific effort that could be reduced if we just sent in dolphins and let them monitor the conditions and give us a report. Give a dolphin a few fish and it will evidently do anything. Dolphins have a wonderfully developed sense of humor, although their idea of a joke is sometimes not very funny to us. They tend to be practical jokers.

If we are going to have dolphins cavorting over the oceans of Mars, then we might as well have horses cavorting over the terraformed landscape. A horse's basic function is to clip the ground cover and produce an amazing amount of not unpleasantly odorous fertilizer; it also happens they are kindly disposed to us and we can ride them as well. They are quick to recognize that humans can get tastier food than they can and so they are easily conned by their two-legged friends.

The sea god Poseidon created the horse, according to the imaginative Greeks. The result was Pegasus, whose wing-loading, if we are to believe the ancient dimensions for the creature, was insufficient for flight — on Earth. But with a one-bar, dense, Earth-like atmosphere and only 38 percent gravity, Pegasus might fly on Mars. He also might exist: gene splicing is to biology what an artist's pen is to the imagination of a comicbook writer. Someday it might be possible to ride — or fly — on Pegasus beneath the still-strange Martian sky.

If we want to remain within something like the bounds of possibility, however, we may have to settle for an ordinary horse. There is a firmly rooted and generally wholesome love for horses in us. They are loyal and trusting and intensely loving when they are treated well; if treated ill, they show their disenchantment with the human race in stubbornness and bad temper. They are much larger than we are, and it is harder to intimidate a reluctant horse than most people think. We still keep them around, despite having practically no use for them any longer, except for pleasure. But we must remember that we once used horses every day.

Cars, if they are ever available on Mars, would have to be made from native materials to be owned by more than a few. The Martians can build roads if they want, made partly of tar or something very much like it. There are asteroids in the belt associated with Uranus and Neptune that may be partly composed of a form of primordial goo. The black stuff, not unlike tar, could be transported to Mars for road building.

But it seems unlikely that Mars will ever sprout freeways, autobahns, or autoroutes across the landscape. There is little point to it and the surface is a bit rough. It is more likely that it will be an airborne society. The hydrazine aircraft that will be so important to developing the Martian colony and the solar-powered dirigible will serve admirably. But will everyone have one? There is plenty of magnesium for building them from native sources, but manufacturing sophisticated machinery such as hydrazine engines on Mars is probably some years away. In any case, building them would be vastly more expensive than ordering a practically weightless biological choice from the *Montgomery Ward's Genetic Catalogue for the Compleat Martian Homestead.* So why not a horse?

Horses, in fact, would be very happy on Mars. The lighter gravity is a blessing for them. The horse is not terribly well designed: its legs are feeble for what it is required to do, even in natural surroundings. On Mars, horses would live to increasingly old ages, survive falls without breaking legs, stand forever on their fingernails without leg problems (a horse's hooves are actually derived from a primordial ancestor's fingernails). They would be as companionable as they have always been, and as willing to work. Besides, they would be great for the rosebushes.

The planet will have water. It will have algae and lichens. The air is fine for horses if it is good enough for humans. A horse is better in rough terrain than almost any mechanical vehicle; that's why ranchers still keep them in spite of their Land-Rovers and Beechcraft airplanes. That and sentiment, perhaps.

What will be used to till the first farms that are laid out in the open beneath the strange moons when the atmosphere is ready for us and our kind of life? Will the Martians have tiny hydrogen peroxide–powered plows? Will they use small tractors with tiny chemical engines that leave a residue of white wherever they go? They probably will. But they could also use horses, as we have used horses or mules for thousands of years. The horses don't seem to mind very much and they can produce a new foal every year or so to keep up with horsepower requirements.

Will Deirdre's son walk someday behind a team of horses on a Martian farm? Will he gather up the spun magnesium traces of a harness and walk the furrows of a soil not terribly different from the soil his great-grandfather might have walked behind a team of horses? Can we not see the colony on Mars expanding and growing into a settlement out on the surface of the planet? Will their lives, ultimately, be so different from ours, despite being on another world?

Mars may be an odd combination of the very technological and the very basic. Frontier environments breed strange people and have unusual requirements. Colonies do not always — or for long — completely resemble the mother society (or perhaps, more accurately, the mother technology) that made them. Yes, I think there will be horses on Mars. I think Deirdre's son will one day sit upon a high hill above a settlement on a gray horse and survey his homestead. He will do it, or his son or daughter will.

While we may leave many of Earth's species just as they are if we transport them to Mars, gene splicing can populate the planet with some extraordinary creatures adapted from those of Earth. Edgar Rice Burroughs's "thoats," the imaginary mounts of Barsoom, could be biologically manufactured if we want them. A cross between a horse and a rhino with the disposition of a camel doesn't seem too interesting, despite the eight legs; but if we're curious to see one, it could be created with synthetic genes and some very fancy manipulations in the biological laboratories of the Red Planet.

What would be more interesting, however, is what gene splicing could do if we ever found an extinct Martian species. On Earth, some important work has been done with our own extinct species. The last guagga munched its way along the South African slopes at least a century ago. The guagga was a sort of zebra: about three million years ago the mountain zebra and the guagga had a common ancestor. One guagga thoughtfully left a piece of muscle tissue in a salt-preserved pelt that has been in a museum in Germany for years. In the muscle tissue is the legacy of the guagga's DNA. Scientists in 1984 obtained gene fragments from the muscle tissue and were able to isolate the DNA and then clone the gene samples.

A mammoth, much more ancient than the guagga, was frozen into oblivion in Russia in a distant ice age. Small pieces of DNA that have been isolated from the frozen mammoth resemble the DNA of elephants, which proves that they are indeed DNA bits from the mammoth, not from something else that may have died with the creature.

The most exciting result of this biological foray into the weird is that intact genes from extinct species that died millions of years ago — genes from muscles, bones, and teeth — have been isolated. It is not yet possible to use gene splicing or any other biological technique to re-create an extinct species, but the directions of genetic research show that it might not be a problem for some twenty-first-century laboratory researcher. We might not just see carefully tended, strung and wired skeletons of the fabulous beasts that once roamed our planet; we might see real ones in a special zoo/museum. The dinosaurs will once again roam the Earth, but in captivity. We can see the odd creatures that were the first birds, can come close to the compact animals that were the most ancient ancestors of those species we recognize today.

And if we someday found an extinct truly Martian animal, what then? Might we not have it again to keep our own animals company? What if one did once exist, if we re-created it and then gene-spliced it with some recognizable Earth species . . .

But this is wandering around in biological quicksand from which we might or might not extricate ourselves. What people take to Mars for company a hundred or more years from now might be beyond even our wildest ramblings; the odds are, however, that it will be a slow road to multiple species on Mars and that Earth's plants and animals will be carefully selected.

Whatever is done will have a distinctly Martian flavor. A forest might exist one day, deep in one of the areas where the atmospheric density is the greatest. It might not be the extensive and carefully cultivated gardens of Versailles or the sprawling, natural forest of Sherwood, but it will be there.

The trees, growing in the lesser Martian gravity, would be thinner and taller than we are used to. The tops would spread out flattened, as are the trees of the South African veldt. Between the reedlike tree trunks would be a carpet of grass, thick, dark green, and luxurious: a forest floor of *O. japonicus,* mondo grass. Each dense clump will spread farther from the nucleus than the same grass clump would on Earth. Here and there, small purple flowers will be buried among the green.

Ivy will be growing on the trees, as will moss; neither of them may be on the north side, as is true on Earth. Heather and Irish heath will be mixed among the mounds and hills of the woodland; their bell-shaped flowers will add color to the landscape. Dichondra will be growing there too, and New Zealand brass buttons, with their delicate, fernlike foliage and the tiny, green-yellow flowerheads that gave the plant its name.

If we can contemplate a forest of this sort someday on a terraformed Mars, we should also have deer for the forest. Perhaps they would be like "Chinese deer," small — about the size of goats — but a quick glance at the rack spread of an angry male would convince anyone they might be small, but not defenseless. Herds of miniature deer of this type roam in the park-sanctuaries in the Midlands of England.

Perhaps in our far Martian future, the animals of the planet will mostly be genetically altered to fit some dictates of the Martians, but we would still recognize them; their familiarity to us is why they would be there in the first place. But the animals, genetically modified or not, may not *remain* familiar to us.

The problem is that gene damage or mutations might take place in the more exposed Martian environment. Mars may never have the great radiation belts that surround Earth and protect us to some degree from harmful radiation. Our Van Allen Belts, as they are called, are primarily a function of our well-developed magnetic field, and so far we do not think Mars has much of a magnetic field.

If Mars does suffer from radiation effects profoundly different from Earth's, the plant and animal life will be affected in the long run. It is impossible to predict with any certainty what the result might be, but it is most likely that some mutations will develop. If that is so, the species on Mars will slowly differ more and more from their cousins on Earth.

While genetic change is hard to guess, what we can predict is that the animals will not need such strong muscles in the weaker gravity, and sooner or later Mother Nature, Martian or not, will recognize this. The result may be longer legs, fewer muscle bunches, and about the same physical abilities. Plants will grow taller than they do here and will look a bit ethereal.

The light on Mars is always less than on Earth, even in bright sunlight. The eyes of animals may adapt to the weaker light conditions by growing to maintain the same acuity. This is true for animals that will remain in the open at some time in the Martian future. Those kept in artificial conditions among humans and human artificial lighting will not change.

Mars will never be as warm as Earth, so animals will be a good deal shaggier than what we are used to, except for those that have developed in Earth's cold climates and are already naturally shaggy. Our anxious horse, for example, which we have not so surreptitiously placed on Mars, might eventually seem rather strange to our Earth eyes. It would be long-haired, almost as much as a long-haired dog.

Its legs would be longer than on Earth (the genetic memory would eventually be convinced that the weaker gravity was a permanent fixture) and its body more sausage-shaped. The chest would be deeper and broader. The great eyes would be almost a quarter larger than normal, liquid and luminous; the third eyelid with which a horse is equipped would remain, but would be larger. The third eyelid acts as a windshield wiper for the eyeball and will be kept busy wiping away the ever-present dust of the Red Planet.

And what of the Martians? Will they change too? Will years upon years and generations upon generations bring genetic changes that will make the people strange to us?

The first change may be taller and, presumably, proportionally thinner-boned humans. The weaker gravity will probably affect human growth in somewhat the same way zero-g does: a stretching of the bones and an inch or two in height on the average. This is hardly certain, however. We have never been able to study physiological changes in humans in a weaker (but not absent) gravity situation for any length of time. The days spent on the moon during the Apollo missions hardly count and were preceded and followed by four-day periods of zero-g, which spoiled the data to some degree.

If our Martians are eventually able to remain outside, they may experience the slight increase in eye size that some of the animals will. Some studies done on Earth suggest that evolutionary changes relating to the amount of sunlight are common among most species, and it is likely that they will happen to us if we live on Mars long enough.

The Martians' Earth muscles will be less useful to them and will tend to atrophy. Children born on Mars will have less use for the bundles of muscle Earth's evolution provided them with, and some of the musculature we are familiar with here may disappear.

If the decrease in muscular ability is serious enough, the Martians may adopt artificial exoskeletons for doing heavy work. Exoskeletons are worn outside the flesh the way a lobster wears its shell. The Martians could use an exoskeleton powered by hydraulic motors as a muscle substitute. Sensors inside the unit would detect any tiny movements of the person wearing it and would translate those movements into signals for the motors. This sort of device has been designed and tested several times already by various militaries. In the American military, it is called a "man amplifier." One design enables a human to lift and support half a ton on either hand.

The logical extension of man amplifiers is the so-called

cyborg, in which a human being is imbedded in a machine and controls the machine by thought alone. Research into cyborg applications has been going on for years, and for obvious purposes, such as piloting fighter planes. Whether a cyborg would be useful on Mars is debatable, but it will probably be available to the Martians eventually.

What will we find as the generations flow by for those on Mars? Besides some possible evolutionary changes in us humans, there will be social changes. The Martians will have their own distinct culture. Whether it will resemble ours or be wildly different is hard to predict. Most colonies end up having a fair resemblance to the mother country; the United States, despite its serious differences in outlook, superficially still resembles England in many ways after two hundred years of independence. Americans are much closer to their British cousins than to the great cultures of Asia, for example.

But Mars may be quite a new direction in human history. For one thing, its population may come from many of Earth's countries via crews at the space stations and it may therefore have a rather chaotic history as it sorts itself out into a defined cultural group. For another, it is an alien environment and will always remain so, whereas colonies on Earth have always been established in relatively familiar surroundings.

What may happen over the years is that the Martians will become the great pioneers of space; they will develop Mars, become miners in the asteroid belt, launch expeditions for the unexplored reaches of the solar system. They may send the first ship to Titan, the great moon of Saturn, that has an atmosphere denser than Mars will have when it is terraformed. Titan's atmosphere is regarded as being a kind of primordial soup in which basic organics (life) might exist.

They may send the first ship to the major outer planets,

Uranus and Neptune, and their moons. The Martians may launch a ship someday into the darkest outlands of our system and visit far Pluto, about which we still know practically nothing except that it may have been a moon of Neptune at one time — and again, perhaps not.

The Martians may pursue astronomy beyond the present boundaries of the solar system. They may someday go in search of a planet "X," the theoretical tenth planet. Scientists have suspected its existence because of slight changes in the orbits of Uranus and Neptune that cannot easily be explained by the gravitational disturbance caused by Pluto, which is quite small and not very dense.

Planet X may be one hundred times farther from the sun than Earth is. It may have a path around our star that takes it more than a thousand years to complete. If it does wander about out there, on the far fringe of our solar system, scientists believe that it may trigger comet showers every twenty-eight million years because of complex gravitational processes.

In the far future, because of their familiarity with space technology and their position in space, the Martians may be the pool from which the crew of the first starship is drawn. Predicting that time is even more difficult than speculating about colonizing Mars. Starships have long been written about, but never perfected. Only a fusion engine of unprecedented efficiency and with a massive weight of fuel taken from the atmosphere of Jupiter could be used for such a craft, as far as any available designs today go. We don't even have fusion in the laboratory yet, so a Martian starship may be a distant dream.

Even if the engine could be built and enough fuel carried, a ship might take more than a century to reach the nearest star, Alpha Centauri. But it may happen. Time and human events are a double helix: given enough time, we can do the most amazing things. Eight thousand years ago, we couldn't even write our own names, let alone describe the behavior of rockets in interplanetary space. Now we have landed on the moon.

Mars is out there waiting. And the stars, too.

EPILOGUE

I WON'T BE on any spaceships to Mars. I will never set foot on those red, alien sands that have so long fascinated me, see the strange pink sky, or watch the hurtling moons perform their intricate dance of phases.

Being the only honorary member of the American Society of Aerospace Pilots does not make me an astronaut; in any event, the people who make the decisions about who goes and who stays do not appoint honorary anythings to do real jobs. I'll be sixty years old when the first ship leaves, according to the timetable of this book.

As we get older, we sense the wonderful changes, the approach of the new, the advantages of that stuff we call technology, and at the same time we feel it receding from our grasp. What was it the poet said? "I want a warm summer's day. I want to ride a Carousel horse; lately I have been lusting after immortality too."

Robert Goddard, who invented the rocket, knew that

some version of it could go to Mars. He knew it as a scientist and engineer knows such things. He did not see the event. He died before von Braun's V-2 rose from the Baltic and swung in its ballistic path toward London. Dr. von Braun, who saw the ship land on Mars as clearly as anyone has, died before his shuttle freighter that could be used to build a Mars ship ever left Cape Canaveral for an orbit above Earth.

It takes so much time, this transferring of theory to fact — more time than any of us think, even in that sometimes most lucid moment just before sleep, when the dreams are often the clearest and yet the most vague. If we could only do *now* what we imagine!

I was a member of the old Moonwatch program while still in high school; I think it was in 1958 or 1959. Probably no one now remembers Moonwatch. We had special binoculars with which to scan the lunar surface; a rocket was going to impact there. The nose cone contained flour, or something like that, and we were to record the moment the flour spread over a tiny part of the moon. For a brief instant in time we on Earth would know that the world would never be the same again — nor would mankind.

From this distance in time Moonwatch seems slightly ludicrous, like someone talking about cranking up the Old Jenny and driving to town at 20 miles per hour over muddy roads. They used to talk about cars the same way they talked about mules, that being the nearest equivalent in their experience. Female mules are called jennies.

If you want to go left with a mule team, you yell "Gee"; if you want to go right, you yell "Haw." This I learned as a very young boy from my grandfather. He didn't use the mules any longer; he had a tractor. But occasionally he waxed strangely nostalgic — and instructive, too. It must have been those times when he thought of hitching up the mules again.

He was terminally fascinated with astronomy. "See those lights up there," he once said, "those are planets. They're places like this, I think. Anyhow some of the scientists think so." He had studied astronomy at Liberty College, an institution now long dead, some years before 1900. He was, regrettably, quite old when I knew him, and I was too young to remember enough.

One of the problems of this culture that quickens us so fast is that we don't listen to our grandfathers. If we did, we might discover what generations instead of years can accomplish. We can see the end of the tunnel, yet we disappear along the way; the Great Leveler takes us out of the game before we see who wins. That's what grandfathers are for: to let us know the scope that generations, not lifetimes, can bring about. Grandfathers understand far better than their sons that we will live on Mars someday.

My grandfather was born into an agricultural nation and lived to see technology take hold: the telephone, the automobile, the Hotchkiss machine gun, the battleship, movies, the airplane, radio, radar, and television. I am sure he never quite imagined that if he had lived long enough he would have seen the surfaces of the planets that so fascinated him on the six o'clock news.

I believe he must have read Lowell. He was in his thirties when Lowell's first book on Mars was published; they were contemporaries. He saw the newspaper clippings of Goddard's rocket projects; he read about von Braun's Peenemunde rockets, the V-2.

From behind a mule team he watched a parade of history that left mankind on the threshold of space. In the year he died, a V-2 with a Wac Corporal sounding rocket as a second stage went to an altitude of 250 miles. I don't remember him talking about space travel, but I suspect he did. I remember the mules only vaguely; my grandfather could summon up the smell of turned furrows, sweaty animal hide in the hot sun, the feel of the cracked leather traces.

The sack of flour that impacted on the moon in the late 1950s was not unlike a sack of flour I once sat on in a rural store, no doubt resting from a hard session with grandfather and the mules. That a sack like that should reach the moon in my lifetime is almost magical. That I would see, via television, the first man set a ribbed boot on the moon is even more so. Now I gleefully talk about living on Mars in great comfort someday.

I could be a passenger on the space shuttle. NASA has been talking about passengers aboard the shuttle for years now. While I must be on the bottom of the list as a candidate, I am also in fairly good health and the shuttle is extremely gentle on people. I've a few years to go before that option is entirely eliminated.

The first expedition to Mars will depart from the station while I am still well enough put together to jump my horse. But I will see it from a distance, via electronics. No one is going to take my then-aging body to the Red Planet. I may watch the great solar sails depart for the Deimos supply dump. The first colony may be established in my lifetime. Beyond that my vision becomes murky.

My daughter may yet become an astronaut. She says she wants to. She could be in the crew of a space station. My daughter could be on the first Mars expedition. She could be Deirdre's mother; Deirdre might be my grandchild. What a thought! That I, who have had a slight acquaintance with driving a mule team, thanks to my grandfather, should become a grandfather who might have gone into space and whose grandchild might be born in orbit around Mars.

And what of Deirdre? Will her schoolbooks have a picture of a mule? More likely they will have pictures of a space scooter; she'll check one out from terminus 354 of the space station and visit friends in other orbital houses. Will she someday tell her own child about the grandfather who wrote about space and was so old he lived before any artificial satellites were even in orbit, before the space station? Before the Mars colony? A sort of Shakespearean figure in funny clothes, glaring from an old photograph. Or perhaps a hologram.

Deirdre's grandchild mounts a horse on a hill above the settlement. Below is something my grandfather would recognize: a struggling community of friends, lovers, fathers, and mothers making their way through time. A church, perhaps, in the town, recognizable to us only by the inevitable cross above the stretched plastic of the orange-segment dome.

There is a community store, the equivalent of flour sacks on the floor. Things don't change much, except that this community and church and store are on a different planet. My daughter's great-grandson living on Mars in much the same way my grandfather did in his rural village — a few more conveniences, but not many. That's all the time we've taken in this journey into the future to come almost full circle. Mules to space stations to Mars. That quickly.

Hell, maybe he'll name a hill on the homestead after me. Even a little one would do.

SUGGESTED READING

THE FOLLOWING LIST of books ranges from 1799 to the present time. Some of the more historical items are available in major libraries or the libraries of most observatories. They are interesting because you can trace the hopes and dreams about Mars as the discoveries with telescopes came in.

The pre-spacecraft period is especially interesting because the conditions on Mars were still somewhat unknown and there was still speculation about the canals and life there.

The modern era of spacecraft has provided the basis for exploration in the next few years and for eventually living on the Red Planet.

THE BEGINNING

Astronomy, James Ferguson. J. Johnson, R. Baldwin, et al., Publishers, London, 1799.
Astronomy, James Ferguson (Brewster 2-vol. edition). G. & W. B. Whittakers, London, 1821.
The Planetary System, J. P. Nichol. H. Bailliere, London, 1850.

La Pluralité des Mondes Habités, Camille Flammarion. Didier et Cie, Paris, 1865.

Les Mondes Imaginaires et les Mondes Réels, Camille Flammarion. Didier et Cie, Paris, 1868.

The Orbs Around Us, Richard A. Proctor. Longmans, Green, and Co., London, 1872.

La Planète Mars, Camille Flammarion. Gauthier-Villars et Fils, Paris, 1892.

Mars, Percival Lowell. Houghton Mifflin Co., Boston, 1895.

Mars and Its Canals, Percival Lowell. Macmillan Co., New York, 1906.

Mars and Its Mystery, Edward S. Morse. Little, Brown and Co., Boston, 1906.

Mars as an Abode of Life, Percival Lowell. Macmillan Co., New York, 1908.

Mars, William H. Pickering. Richard G. Badger, Boston, 1921.

THE MIDDLE PERIOD

Biography of Percival Lowell, A. Lawrence Lowell. Macmillan Co., New York, 1935.

Observations of Mars and Its Canals, Harold B. Webb. Privately printed at Jamaica, New York, 1936 and 1941.

The Lowells and Their Seven Worlds, Ferris Greenslet. Houghton Mifflin Co., Boston, 1946.

The Planet Mars, Gerard de Vaucouleurs. Faber and Faber, Ltd., London, 1952.

The Green and Red Planet, Hubertus Strughold. University of New Mexico Press, Albuquerque, 1953.

Exploring Mars, Robert S. Richardson. McGraw-Hill Book Co., New York, 1954.

Physics of the Planet Mars, Gerard de Vaucouleurs. Faber and Faber, Ltd., London, 1954.

A Space Traveler's Guide to Mars, Dr. I. M. Levitt. Henry Holt and Co., New York, 1956.

The Exploration of Mars, Willy Ley and Wernher von Braun. Viking Press, New York, 1956.

There Is Life on Mars, The Earl Nelson. Citadel Press, New York, 1956.

The Mars Project, Wernher von Braun. Illini Books, University of Illinois, Urbana, 1962.

Mars, Earl C. Slipher. Sky Publishing Corp., Cambridge, Mass., 1962.

Mars, Robert S. Richardson. Harcourt, Brace & World, New York, 1964.

The Book of Mars, Samuel Glasstone. NASA SP-179, U.S. Government Printing Office, Washington, D.C., 1968.

THE SPACECRAFT ERA

The Book of Mars: Tales of Mars, Men and Martians, ed. Jane Hipolito and Willis E. McNelly. Futura Publications Ltd., An Orbit Book, 1976. Originally *Mars, We Love You,* published by Doubleday & Co., 1971.

The Mariner 6 and 7 Pictures of Mars, Stewart A. Collins. NASA SP-263, U.S. Government Printing Office, Washington, D.C., 1971.

Mars and the Mind of Man, Ray Bradbury, Arthur C. Clarke, Bruce Murray, Carl Sagan, and Walter Sullivan. Harper & Row, New York, 1973.

Mars as Viewed by Mariner 9. NASA SP-329, U.S. Government Printing Office, Washington, D.C., 1974.

The New Mars. NASA SP-337, U.S. Government Printing Office, Washington, D.C., 1974.

The Viking Mission to Mars, Robert M. Powers et al. LC-75-16764, Martin Marietta Corp., Denver, 1975.

The Next Fifty Years in Space, Patrick Moore. William Luscombe, London, 1976.

From Vineland to Mars, Richard S. Lewis. New York Times Book Co., New York, 1976.

Lowell and Mars, William Graves Hoyt. University of Arizona Press, Tucson, 1976.

The Geology of Mars, Thomas A. Mutch, Raymond E. Arvidson, James W. Head III, Kenneth L. Jones, and R. Stephen Saunders. Princeton University Press, Princeton, N.J., 1976.

The Solar Planets, V. A. Firsoff. Crane, Russak & Co., New York; David & Charles, London, 1977.

Guide to Mars, Patrick Moore. W. W. Norton & Co., New York, 1977.

SUGGESTED READING

Mars and Its Satellites, Jurgen Blunck. Exposition Press, Hicksville, N.Y., 1977.

Mars, the Red Planet, Isaac Asimov. William Morrow & Co., New York, 1977.

Viking Lander System. Primary Mission Performance Report, NASA CR-145148, Martin Marietta Corp., Denver, 1977.

Exploration of the Solar System, William J. Kaufmann III. Macmillan Publishing Co., New York, 1978.

Mars, Jeff Rovin. Corwin Books, Los Angeles, 1978.

Cesta na Mars, Karel Pacner and Antonin Vitek. Albatros Publishers, Prague, 1979.

Life on Mars, David L. Chandler. E. P. Dutton, New York, 1979.

Shuttle: World's First Spaceship, Robert M. Powers. Stackpole Books, Harrisburg, Pa., 1979.

Toward Distant Suns, T. A. Heppenheimer. Stackpole Books, Harrisburg, Pa., 1979.

Space Resources and Space Settlements. NASA SP-428, U.S. Government Printing Office, Washington, D.C., 1979.

Solar System, Peter Ryan and Ludek Pesek. Viking Press, New York, 1979.

Atlas of Mars, R. M. Batson, P. M. Bridges, J. L. Inge. NASA NAS 1.21:438, U.S. Government Printing Office, Washington, D.C., 1979.

The New Face of Mars, V. A. Firsoff. Ian Henry Publications, Ltd., Essex, Eng., 1980.

The Search for Life on Mars, Henry S. F. Cooper, Jr. Holt, Rinehart and Winston, New York, 1980.

The Fertile Stars, Brian O'Leary. Everest House, New York, 1981.

New Earths, James Edward Oberg. Stackpole Books, Harrisburg, Pa., 1981.

The New Solar System, ed. J. Kelly Beatty, Brian O'Leary, and Andrew Chaikin. Sky Publishing Corp., Cambridge, Mass., 1981.

Understanding DNA and Gene Cloning, Karl Drlica. John Wiley & Sons, 1984.

Mission to Mars, James E. Oberg. Stackpole Books, Harrisburg, Pa., 1982.

The Channels of Mars, Victor R. Baker. University of Texas Press, Austin, 1982.

Planets of Rock and Ice, Clark R. Chapman. Charles Scribner's Sons, New York, 1977, 1982.

Planetary Encounters, Robert M. Powers. Sidgwick & Jackson, Ltd., London, 1982.

Travelers in Space and Time, Patrick Moore. Doubleday & Co., New York, 1983.

Project Space Station, Brian O'Leary. Stackpole Books, Harrisburg, Pa., 1983.

On Mars: Exploration of the Red Planet. NASA SP-4212, U.S. Government Printing Office, Washington, D.C., 1984.

The Case for Mars, ed. Penelope J. Boston. American Astronautical Society, San Diego, Cal., 1984.

The Case for Mars II, ed. Christopher P. McKay. American Astronautical Society, San Diego, Cal., 1985.

INDEX

INDEX

meteors, 161; *see also* asteroids
methane, 73
microorganisms, 18–19, 23, 82
Milky Way, 40
mineral resources: from Mars, 94–96; from space, 159–64
minute on Mars, 68
mirrors: on spaceship, 55; for terraforming, 149–51, 154–55
month on Mars, 87
moon of Earth, 12, 137, 188; landing on, 36, 44; mineral resources of, 159–61; ownership of, 130
moons of Mars, 70–71, 79, 86–94, 177; description of, 88; eclipses by, 89; rotation of, 88–89; size of, 88; terraforming, 188–89; as way stations for colonization, 89; *see also* Deimos; Phobos
Moon Treaty, 130
Moonwatch, 214
movies about Mars, 5, 6
muscle atrophy on Mars, 209; *see also* exercise
mutations. *See* Genetic adjustments and mutations
Mutch, Thomas A., 21; *see also* Thomas A. Mutch Memorial Station
Mylar, 73, 150
myths, 10

NASA, 12, 21, 36–38, 100, 140, 143, 162, 185, 216; geochemical orbiter for Mars, 54; and manned spacecraft to Mars, 45; unmanned robot probes, 93; *see also specific programs of*
natural resources: Earth, 159; Mars, 94–96; moon, 159–60; space, 159–64
navigation on Mars, 97
Nelson, Earl, 195
Neptune, 15, 50, 211
Nessonis Lacus, 67
Newton, Sir Isaac, 44
Newton's Third Law, 60
nitrogen, 17, 73, 176–77
nuclear reactors' use on Mars, 99

Oberth, Hermann, 149
oceans: Earth, 184–85, 187; possibilities on Mars, 184–87
oil sources, 160

Olympus Mons, 7, 17, 54, 65, 79, 107, 181–82
orbital tilt of Mars, 142
orbiters: from Deimos, 92
orbiting space stations. *See* space station
Outer Space Treaty of 1967, 129–30
oxygen, 8, 12, 13, 17, 18, 96, 170, 171, 173, 175, 176; and colonization, 24–25; sources of, on Mars, 72, 73, 152; and walking, 76; *see also* atmosphere of Mars

parthogenesis, 199
patera, 182
Pavonis Mons, 65
penetrator probes, 92
permafrost, 22–23, 72, 78, 95, 112, 144, 151, 152, 154
Phobos, 7, 67, 86, 143, 177; description of, 87, 88–89; eclipses by, 89; a "terminal" moon, 89; terraforming, 154–55
Phoenicis Lacus, 78
phosphorus, 96
photography of Mars, 2, 16–21, 88, 105–6, 107; see also *Viking* program
photons, 90–91
photosynthesis, 170–72
pigs, 103–4
Pioneer, 61
plague bacilli, 191–92
planetary engineering, 134; *see also* terraforming
planetary system formation, 135–37
Planet Mars, The (Flammarion), 29
planet X, 211
plants on Mars. *See* agriculture on Mars; forests on Mars
plastics, 121
Pluto, 211
polar caps of Mars, 12–13, 23, 28, 95; melting, 151, 153–54; and terraforming, 146
pollution of space, 111
power supply: for colonization, 99; for rovers, 104–5
P. pestis, 192
pregnancy aboard spaceship, 62–63
press coverage of Mars expedition, 82
Princess of Mars (novel by Burroughs), 31–32